T0200121

PHILOSOPHICAL METHODOLOGY

PHILOSOPHICAL METHODOLOGY

From Data to Theory

John Bengson

Terence Cuneo

Russ Shafer-Landau

OXFORD

UNIVERSITY PRESS

OXFORD
UNIVERSITY PRESS

Great Clarendon Street, Oxford, OX2 6DP,
United Kingdom

Oxford University Press is a department of the University of Oxford.
It furthers the University's objective of excellence in research, scholarship,
and education by publishing worldwide. Oxford is a registered trade mark of
Oxford University Press in the UK and in certain other countries

First Edition published in 2022

Impression: 2

Published in the United States of America by Oxford University Press
198 Madison Avenue, New York, NY 10016, United States of America

British Library Cataloguing in Publication Data
Data available

Library of Congress Control Number: 2021951475

ISBN 978-0-19-286246-4 (hbk.)
ISBN 978-0-19-286247-1 (pbk.)

DOI: 10.1093/oso/9780192862464.001.0001

Printed and bound by
CPI Group (UK) Ltd, Croydon, CR0 4YY

CONTENTS

ACKNOWLEDGMENTS

This book has a somewhat unusual origin story. In 2013, the three of us hatched a plan to write a book in metaethics, an early version of which was the topic of a three-day workshop at the Vrije Universiteit, Amsterdam in 2015. That workshop transformed our project. One important critique went something like this: "You've advertised your book as implementing a distinctive philosophical methodology. But you've only sketched its contours. Let's hear more!"

Attempting to respect this admonition proved to be challenging. The more we thought about philosophical methodology, the more we felt needed saying. So rather than simply fold our views on this topic into a manuscript on moral philosophy, we determined to write a self-standing monograph on philosophical inquiry in general. The result is the short book before you.

Our debts to those who helped us are many. We're especially grateful to René van Woudenberg and his colleagues for their contributions to the workshop they organized at the Vrije, and to Selim Berker, Anandi Hattiangadi, Tristram McPherson, and Mike Ridge for having taken the time to read our work and provide extensive and penetrating criticisms of it at this event. Thanks also to Peter Momtchiloff, our stellar editor at OUP, who enlisted Chris Daly, Joel Pust, and (once more) Tristram McPherson to serve as reviewers; each offered helpful feedback on a draft of the manuscript. Addressing their incisive comments extended our project by many months—all, we hope, in the aid of a work that is clearer and deeper than it would've been without the benefit of their critical attention.

We initially aired some of our ideas on methodology in a pair of articles from 2019,[1] as well as at numerous venues, including Brown University, Jilin University, the German Society of Analytic Philosophy in Berlin, Lingnan University, the LOGOS Research Group at the University of Barcelona, Harvard University, MIT, Stockholm University, the University of Vermont, the University of Wisconsin–Madison, Washington University in St. Louis, and the University of Texas at Austin. Thanks very much to all of the participants at these colloquia, conferences, and seminars, as well as to many other generous interlocutors, for illuminating discussions that greatly improved our work.

---

[1] Bengson, Cuneo, and Shafer-Landau (2019a and 2019b).

LIST OF ILLUSTRATIONS

Introduction

This book is an attempt to understand philosophical inquiry. We suspect that you, like us, have grappled with a philosophical question—perhaps a question about consciousness, or right action, or free will, or the possibility of knowledge. You've felt its force, as well as the urge to understand, to make sense of whatever the question is about. And so commences a familiar philosophical venture: to find, or develop, a theory that succeeds in doing just this.

We are motivated to take up this topic for several reasons, beyond our conviction in the value of understanding what we're doing when we do philosophy. One is that commitments regarding inquiry deeply shape philosophical discussion and debate. When these commitments diverge, as they often do, this can lead to serious misunderstandings and incite disagreements downstream—a "battlefield of endless controversies," in Kant's phrase.[1]

---

[1] Kant (1781/1998, Aviii).

Philosophical Methodology: From Data to Theory. John Bengson, Terence Cuneo, and Russ Shafer-Landau, Oxford University Press. © John Bengson, Terence Cuneo, and Russ Shafer-Landau 2022. DOI: 10.1093/oso/9780192862464.003.0001

And yet these effects are avoidable. Just as Buddhism teaches that a certain kind of insight eliminates the source of suffering, we're convinced that a clearer view of philosophical inquiry is apt to help dissolve misunderstandings and disagreements whose sources are methodological.

Another reason to write this book, which takes on special importance when viewed against these observations, is to equip philosophers to satisfy the imperative to be transparent and explicit about their methodological commitments. Unfortunately, this imperative is nowadays honored largely in the breach, leading some to speak of a methodological "scandal" or "crisis," and to express "hope" for a sort of "methodological consciousness-raising."[2] Such talk may be hyperbole. But even if it is, the problem is very much with us. Peter Godfrey-Smith isn't far wide of the mark when he writes,

All of philosophy is plagued with discussion and anxiety about how philosophical work should be done and what a philosophical theory should try to do.[3]

At the same time, many philosophers are suspicious of offering anything like a recipe for doing philosophy, much less for doing it

---

[2] The first of the quoted expressions is from Dummett (1978, 458), the second from Collingwood (1933, 4–6; *cp.* Baz 2017), and the final pair from Williamson (2007, 287), who elsewhere speaks of a "slow-burning crisis of philosophical method" (2018, 4). Much philosophy in the twentieth century occurred in the grip (or shadow) of "the linguistic turn" (see especially the essays in Rorty 1967). Philosophy has traveled a long way since, but without a rigorous articulation of a non-linguistic approach to philosophical inquiry.

[3] Godfrey-Smith (2003, 5). Wilson (2017, 91) claims that "the absence of fixed standards in philosophy" leads to various disciplinary problems.

well. Philosophy, it may be said, is too difficult, creative, and unpredictable for there to be a helpful way of describing how it ought to unfold. So we shouldn't aim to devise a theory of philosophical inquiry. Nor should we be advising other philosophers on how to go about their business.

We agree that the search for a recipe for philosophical success is misguided, and that issuing recommendations risks immodesty. But it doesn't follow that there is nothing interesting or important to be said about philosophical inquiry.

For one thing, we can distinguish different types of philosophical inquiry. Philosophy can be undertaken in a practical mode, for example, as a means to some personal, social, political, or ethical end.[4] It is also possible to philosophize in an aesthetic mode; perhaps this is the right way to think of those who engage in philosophical activity just because they find it elegant, arresting, or awe-inspiring.[5] Still other modes are possible. While we ourselves are pluralists, endorsing the compatibility and legitimacy of these different options, we focus here on philosophical engagement in the theoretical mode. As will emerge, we understand this to be inquiry that is aimed at securing an epistemically satisfactory resolution of the questions that prompt it.

---

[4] Therapeutic missions are often read into Socrates' connection between philosophy and "training for death," and attributed to Confucius, the Stoics and Epicureans, Kierkegaard, and the late Wittgenstein, while a political one is gleaned from Marx's eleventh thesis on Feuerbach: "Philosophers have hitherto only interpreted the world in various ways; the point is to change it." *Cp.* Hadot (1981/1995) and hooks (1991).

[5] *Cp.* Nozick (1981, 20).

Our primary goal is to address five central questions about philosophical inquiry, understood as a species of theoretical inquiry. The first two can be stated simply:

> *What is the structure of philosophical inquiry?*
> *What is the principal goal of philosophical inquiry?*

Answering these questions requires developing a model of such inquiry that identifies its basic components, how they're organized, and what those elements, so organized, aim to achieve. We do these things in Chapter 1.

According to the model we develop there, the two main components of philosophical inquiry center on *data* and a *method of theorizing*—the former being starting points of inquiry, and the latter a set of criteria for theory construction and evaluation. The principal goal, we'll argue, is a certain type of *understanding*, which can serve as a measure of philosophical progress. Accordingly, our next three questions are:

> *What are philosophical data?*
> *What is a sound method for philosophical theorizing?*
> *Is there reason to think that philosophical progress can be or has been made?*

The remainder of the book, Chapters 2 through 6, takes up these questions.

We answer the third question, concerning data, by detailing an epistemic theory of data that privileges pre-theoretical claims that inquirers have good reason to believe. Our discussion highlights the role of data in inquiry, defends philosophical data against the charge of theory-ladenness, illustrates the varieties of data and procedures for data collection, and contrasts our preferred view with rivals that analyze data in sociological, metaphysical, psychological, or linguistic terms. We do this work in Chapters 2 and 3.

Our answer to the fourth question, concerning method, revolves around a specific set of criteria for theory construction and evaluation. We'll argue that a theory that satisfies these criteria resolves inquiry by providing a certain type of understanding. In arguing for this claim, our project makes explicit some of its normative commitments: we do not merely offer an account of what philosophical inquiry is, but take a stand on what it is to do it well, by charting how its success is possible. This enterprise takes up Chapters 4 and 5. In Chapter 4, we review some classic and contemporary philosophical texts with an eye to identifying a variety of methodological virtues. We then consider some prominent philosophical methods, pinpointing both their attractions and their shortcomings. This sets the stage, in Chapter 5, for our presentation and initial defense of the philosophical method that we endorse, which we label the 'Tri-Level Method.'

The fifth question, concerning progress, forces us to confront a familiar anxiety about philosophy—namely, that stable success is largely or entirely beyond reach, despite millennia of efforts. In Chapter 6 we attempt to soothe this worry by using the Tri-Level Method to illustrate and explain philosophical progress. Moreover, we introduce a previously overlooked distinction that allows us to block an invidious comparison between advances in philosophy and in other disciplines.

Philosophical questions about inquiry, data, and theorizing are often grouped together, along with other questions, under the heading of 'methodology.' We follow this practice here. There has been no shortage of work on issues under this heading. Indeed, we see our project as contributing to a rich history of reflection on philosophical methodology—hence the title of this book.

However, many of the questions that we've found it important to address when thinking about philosophical inquiry have received relatively little attention, especially in recent debates about methodology. There has been a lot of discussion of questions regarding (for example) the method of cases, the role of intuitions in philosophy, the relation between conceivability and possibility, the status of common sense, the prospects of experimental philosophy, metametaphysics, conceptual engineering, and disputes between naturalists and their opponents—not our primary topics.[6] Comparatively few philosophers working on methodology have attempted to develop models of the structure of inquiry, or proposed accounts of philosophical data, or explicitly stated a set of criteria for philosophical theory construction and evaluation. These are the central topics of this book.[7]

At center stage is the notion of a philosophical theory, view, or position (we'll use these terms interchangeably). We won't attempt to construct a comprehensive theory of theories. But we do recognize the following necessary condition: a theory is a set of claims whose success depends on its ability to adequately handle the data. As we discuss in Chapter 4, there is substantial controversy about what it is to handle the data. Rather than enter the fray at

---

[6] See, e.g., Gendler and Hawthorne (2002), Chalmers, Manley, and Wasserman (2009), Haug (2013), Overgaard, Gilbert, and Burwood (2013), Daly (2015), Livengood and Sytsma (2015), Cappelen, Gendler, and Hawthorne (2016, Part III), D'Oro and Overgaard (2017), Machery (2017), and Burgess, Cappelen, and Plunkett (2020).

[7] Some of the other topics we'll tackle have received comparatively more attention: for example, the status of reflective equilibrium, what we can expect philosophical argument and analysis to achieve, and whether simplicity and other putative theoretical virtues are probative in a philosophical context.

this point, we instead want to distinguish a focal element of our approach (an emphasis on philosophical theory) from a very common alternative—that of providing arguments for one's views and against those of one's opponents.

While theories and arguments interact in various ways, the two are importantly different. Among other things, they have different success conditions. As just noted, theories are successful only when they handle the data. Arguments are successful only when they are sound or cogent, or perhaps persuasive or reasonable. Arguments can be successful even if they or their conclusions fail to address, let alone handle, any data. (Think of an argument for a simple negation—for example, that there are no abstract entities. Such a negation may open the door to further theorizing, but is not itself the provision of a metaphysical theory that handles data about what there is.) Since arguments differ from theories in this way, we shouldn't expect a treatment of argumentation to tell us much about theorizing. Yet many philosophers focus their methodological attention and ambitions almost exclusively on argumentation. To some extent, this is understandable: arguments are important. But theorizing is, too—in fact, we'll argue that a certain type of theory is the principal goal of philosophical inquiry. One unfortunate result of philosophers' zeal for arguments is that philosophical theorizing has not received the attention it deserves. Part of our project is to fill this lacuna in treatments of the methodology of philosophy.

While our discussion covers a fair amount of territory, we have aimed at concision. This book is short. Its primary goal is to present an account of philosophical methodology without chasing down all of the implications of the account or discussing the full variety of issues that bear upon it. In particular, we'll be introducing elements in our account that are susceptible to precisification in different ways, and we'll deliberately present these elements at a level of abstraction that does not take a stand on every detail, but allows for healthy debate about how best to fill them in.

While we will focus on philosophical inquiry, we believe that a good deal of what's to come can be extended to other genres of theoretical activity, such as science, history, and some subfields of mathematics.[8] But we won't always attempt to make these connections or comment on their viability. Nor will we endeavor to canvass all the tools in the philosopher's toolkit. For example, we won't be scrutinizing specific devices or tactics that may be employed in particular circumstances: counterexamples, thought experiments, deductive and inductive inferences, reductio or transcendental arguments, genealogical critique, narrative study, post-modal metaphysical notions, and so on.[9] Nor will we propose heuristics that budding philosophers might use when seeking to up their game.[10] We also do not discuss particular methods associated with specific schools or traditions, such as phenomenology, ordinary language philosophy, experimental philosophy, feminism, or deconstructionism.[11] This book is neither a manual nor an encyclopedia.

It also does not purport to fully cover the complex forms of inquiry undertaken by historians of philosophy. For example, the methodology we'll offer is completely silent on the norms of

---

[8] For recent discussion of what distinguishes philosophy from other fields, see, e.g., Bealer (1996a, 2–3), Deleuze and Guattari (1996), Priest (2006, §4), and the papers in Ragland and Heidt (2001).

[9] Contemporary books of this sort include Baggini and Fosl (2010), Daly (2010), Papineau (2012), and Dennett (2013).

[10] See, e.g., Hájek (2016).

[11] See, e.g., Part II of Cappelen, Gendler, and Hawthorne (2016).

exegesis, which are central to historically oriented philosophical projects. Still, insofar as the history of philosophy is philosophy, as we believe, some aspects of those projects do fall within our purview here.

Although we will often discuss how philosophy is or has been conducted, our project is not primarily concerned to characterize the lived experience of philosophizing or examine the most recent innovations in philosophical inquiry. Such inquiry, understood as a species of theoretical inquiry, should aspire to rise above trends and convictions that happen to be prominent at a specific historical moment or in a particular sociological context. Good methodology should reflect this point, and should indicate how the aspiration can be realized. Moreover, when it comes to philosophical theorizing, it's not the case that anything goes; good theorizing ought to be guided by a sound method. These are among the ambitions we've set for the methodology we'll develop in the following chapters.

There is a sense in which the methodology we defend over the course of this book is novel. At the same time, it doesn't herald a radical reinvention of philosophy. Our primary aims are to advance a model of philosophical inquiry, develop a theory of philosophical data, and assemble a method for philosophical theorizing that, when followed, promotes understanding. The criteria we'll endorse are familiar from the way many philosophers ply their trade, though these criteria have not yet been sufficiently justified, ordered, and integrated in a way that reveals how they can facilitate the principal aim of inquiry. The project is not to inspire a revolution, but to sharpen our understanding of philosophical inquiry, and of how to do it well.

1

The Nature of Inquiry

Aristotle famously wrote that philosophy begins in wonder. Philosophical inquiry is, of course, prompted by much more than this. Curiosity, puzzlement, annoyance, distress, and even deep anger can prompt us to it. Although these states are highly diverse, they have this much in common: they induce us to raise *questions* that we want our inquiry to resolve.

In philosophy, the central questions concern such topics as the nature of reality, the character and extent of our cognitive powers, and the content and status of practical norms. These questions include, but are certainly not limited to, the following:

> *Why is there something rather than nothing?*
>
> *Does God exist?*
>
> *Do we have free will?*
>
> *How is the mind related to the body?*
>
> *Is there such a thing as human nature?*

Philosophical Methodology: From Data to Theory. John Bengson, Terence Cuneo, and Russ Shafer-Landau, Oxford University Press. © John Bengson, Terence Cuneo, and Russ Shafer-Landau 2022. DOI: 10.1093/oso/9780192862464.003.0002

How is knowledge of the external world possible, if it is?

How do thought and language refer to the world, if they do?

What is the nature of meaning?

Are there objective normative standards?

Is there a single, fundamental principle of morality, and if so, what is it?

What is it to live well?

Naturally, philosophers also wrestle with a welter of more circumscribed questions, such as:

Is existence a property?

Which abilities are required for omnipotence?

Does free will imply the ability to do otherwise?

What is belief?

Is gender a biological category?

What is the difference between justified true belief and knowledge?

Is it possible to refer to abstract objects?

Is meaning normative?

What is it for a standard to be objective?

What distinguishes moral principles from merely prudential ones?

Does living well require being virtuous?

To engage in philosophical inquiry is to endeavor to address questions such as these, and those yet more fine-grained. This means treating such questions as open, aiming to respond to them in as satisfying a manner as possible. Ideally, the culmination will be the successful resolution of those questions, the closing of inquiry.

Our focus is on philosophical inquiry in its theoretical mode. Such inquiry is a species of a genus—theoretical inquiry. As such, it will help us understand our target by gaining a clearer view of the genus itself. We'll do this by tackling two fundamental questions about such inquiry. The first concerns its *structure*: what are the basic elements of, or stages in, this enterprise? While sometimes a messy business in practice, we identify two common threads in diverse instances of theoretical inquiry. Our second question focuses on the *goals* of such inquiry: what is its point or purpose? We distinguish different types of goals, arguing that successful theoretical inquiry terminates in a certain kind of understanding.

1. The Structure of Inquiry

Inquiry comes in a variety of forms, each involving a process or activity aimed at some end or goal. In practical inquiry, we engage in deliberation aimed at the performance of an action. In theoretical inquiry, we engage in reflection aimed at the provision of a theory. We'll examine the goals of theoretical inquiry in detail in the next section. In this section, we analyze its structure.[1]

In our view, theoretical inquiry is best modeled as a process structured around two stages. We call the first the *data collection* stage, since it centers on the gathering of considerations (the data) that must somehow be handled when addressing a set of questions with which an inquiry is concerned. We call the second the *theorizing* stage, since it is designed to address those questions by developing theories that handle the data.

---

[1] Some maintain that inquiry is best analyzed in terms of a distinctive attitude or aim (see, e.g., Friedman 2017, 302–3 and Smith 2020, 184). We remain neutral on this issue, providing an account not of inquiry itself but rather of its structure (its components and their organization). That said, our account of this structure might be parlayed into a novel analysis of inquiry.

Each stage has three main elements. Both data collection and theorizing have *inputs*, *outputs*, and a *procedure* or *method*, which takes one from the former (input) to the latter (output).[2] The two stages are intimately connected insofar as the outputs of the first stage are the inputs to the second. Together, when all goes well, they constitute the transition from the opening of inquiry—concerning a set of questions—to its closing. Inquiry concludes with the provision of a *theory*, a set of claims that inquirers advance in the course of satisfying the method's criteria in the second stage.

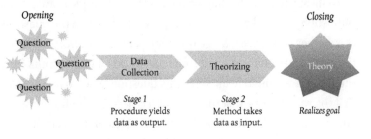

Opening

Question

Question   Data Collection   Theorizing   Theory   *Closing*

Question

Stage 1
Procedure yields data as output.

Stage 2
Method takes data as input.

Realizes goal

The Structure of Theoretical Inquiry

We'll examine each stage of inquiry in detail in subsequent chapters. Here we note simply that this two-stage model is intended to be a theory of the basic *structure* of theoretical inquiry, as distinct from a description of such inquiry as it occurs on the ground. The differences between the two are real: the latter is a temporally extended process that agents carry out more or less competently, while the former is not. Though these differences exist, the structure of theoretical inquiry is hardly divorced from actual inquiries. The activities in which inquirers engage approximate the structure

---

[2] To regiment terminology, we'll be speaking of 'procedures' in the case of data collection, and 'methods' in the case of theorizing.

of theoretical inquiry (or its elements) to different degrees and can be evaluated by the extent to which they conform to this structure. While our model will find a home for the many legitimate things that inquirers actually do, it is not our intention to fully characterize all the workings and difficulties of actual instances of theoretical inquiry, but rather to identify its basic components and to specify their configuration.[3] (Hereafter, as in the title of this chapter, we will often speak simply of 'inquiry,' eliding 'theoretical' for the sake of brevity.)

To illustrate our two-stage model with an example drawn from philosophy, consider a group of inquirers focused on the question of whether God exists. Stage one represents their efforts to identify considerations that must somehow be taken into account by a theory that adequately answers this question. Such data include: that the world exists, that many natural objects are highly complex and organized, that the world contains a tremendous amount of suffering, that numerous individuals have sincerely claimed to experience God, that a great many people report having failed to do so even after making sincere attempts, and so on. Stage two represents the efforts of these inquirers—theists and non-theists alike—to develop theories that account for these and other data. If their theories are good enough, this would thereby close the question that opened the inquiry.

---

[3] Our model allows that inquirers may inquire without a fully determinate question in mind, or in pursuit of an open-ended question such as "What is such and such like?" Inquiries can also overlap with, nest in, and compose other inquiries. Indeed, thinkers often undertake multiple inquiries at once, which combine to form a single, evolving inquiry. For example, investigations regarding defeat, context, and stakes may target questions uncovered in the course of investigating knowledge and justification; all of these inquiries (and others) may be components of one and the same inquiry into skepticism.

Philosophers sometimes flirt with models of theoretical inquiry that do not mark any of its stages. For example, on Robert Stalnaker's familiar and influential picture of "the abstract structure of inquiry," inquirers are treated as believers seeking to narrow the space of possible worlds to a certain subset, one containing those and only those worlds that answer the queries that prompted inquiry.[4] This picture looks seriously incomplete. For it does not tell us what it is for inquiry to unfold; it merely observes that it does. When analyzing the structure of inquiry, what is wanted is insight into the components of inquiry, and how those components are related so as to constitute a certain kind of process. Whatever its merits, the possible worlds picture of inquiry does not provide such insight.

The solution isn't simply to posit multiple stages. Consider, for example, a five-stage view of inquiry advanced by John Dewey, which comprises

(i) a felt difficulty, (ii) its location and definition, (iii) suggestion of a possible solution, (iv) development by reasoning of the bearing of the suggestions, (v) further observation and experiment leading to its acceptance or rejection.[5]

But a process culminating in observation and experiment is insufficient to capture inquiries regarding a wide range of theoretical questions. We aren't going to solve the Liar paradox, figure out why there is something rather than nothing, or ascertain the distribution of primes through observation and experiment—especially if these are empirical undertakings, as Dewey insists.

---

[4] Stalnaker (1984, ix). We'll waive worries about how this picture could account for beliefs regarding specific necessary and essential truths.

[5] Dewey (1910, 72; *cp.* 1938, Ch. VI).

A different sort of problem afflicts the model of inquiry made famous in Descartes' unpublished treatise, *Rules for the Direction of the Mind*. Consider, in particular, Rules 4 and 5:

We need a method if we are to investigate the truth of things.

The whole method consists entirely in the ordering and arranging of the objects on which we must concentrate our mind's eye if we are to discover some truth. We shall be following this method exactly if we *first* reduce complicated and obscure propositions step by step to simpler ones, and *then*, starting with the intuition of the simplest ones of all, try to ascend through the same steps to a knowledge of the rest.[6]

This model posits a structure with two main stages. However, this structure is of the wrong sort, being narrowly focused on executing the Cartesian project of rendering clear and distinct a series of individual ideas or propositions. While this is perhaps one type of theoretical inquiry, it is plainly not the only game in town. Inquiries across the sciences and humanities do not drill down to what is "simplest" and then, using intuition, "ascend" from there to "a knowledge of the rest." A satisfactory account of the structure of inquiry should be perfectly general, applying to every token of the type.

More promising than the possible worlds, Deweyan, and Cartesian approaches is a dialectical model, according to which inquiry consists in formulating and clarifying candidate answers

---

[6] Descartes (1628/1985; emphases added). Later, in the *Discourse on Method*, Descartes identifies four components to his favored method: evidence (assent to nothing that can be doubted), division (divide a problem into parts), order (begin with the simplest parts), and exhaustion (omit no parts). The worries we raise about Descartes' model also apply to models confined to the discovery of necessary or essential truths (as in the Socratic Picture of inquiry described by Bronstein 2016, Ch. 8).

to the questions that open inquiry, and then developing and assessing arguments for and against those answers.[7] This model posits a structure while seemingly avoiding the parochialism of the previous two models. But this alternative does not fairly represent how inquiry unfolds, or what it aims to uncover. When inquiring about a murder, Sherlock Holmes does not simply list suspects—candidate answers to the question "Who did it?"—and then assemble arguments supporting their guilt or innocence. Rather, as the detective makes clear in *A Study in Scarlet*, his inquiry begins with the collection of data:

"You don't seem to give much thought to the matter in hand," I [Watson] said at last, interrupting Holmes' musical disquisition.

"No data yet," he answered. "It is a capital mistake to theorize before you have [them]."[8]

Only after collecting data does the detective pursue a view of the murder that resolves his query. Such a view is more than just the proposition that a certain suspect is guilty. Holmes might achieve a knockdown argument whose conclusion is that Professor Moriarty committed the crime in the drawing room with a vial of poison, but still have no clue why Moriarty was moved to homicide, how he did it with that particular substance, or the extent to which this crime fits with Moriarty's past behavior. In these ways, a dialectical treatment of Holmes' criminal inquiry

---

[7] Different versions of the dialectical model may introduce important details, such as the role of disputation, questioning, or juxtaposition of contradictory answers. See, e.g., Adler (1927, vi), Bird (1953), Gadamer (1960/1989), Mueller (1965), Rosen (1982), Hintikka (1999), and Rescher (2006).

[8] Doyle (1887, Pt. I, Ch. III). In *The Adventure of the Copper Beeches*, Holmes exclaims: "Data! Data! Data!...I can't make bricks without clay."

" THE MAID SHOWED US THE BOOTS. "

Illustration 1 Sherlock Holmes collecting data.

omits important dimensions of his venture. Likewise, a dialectical treatment of theoretical inquiry misses or obscures the role of data, which are fundamental to inquiries in a broad range of fields—biology, economics, and philosophy of religion (recall our example above), to name a few. The dialectical model also distorts the project of constructing a theory that handles the data. The latter goes beyond the mere provision of argument: as we noted in the Introduction, the conclusion of an argument does not a theory make. So, although the dialectical model describes an undertaking that may play a role in theoretical inquiry, it is certainly not the whole story about that process.

This brings us back to the model we're proposing. There are several reasons to prefer a model that incorporates the two stages of inquiry (data collection and theorizing) that we've distinguished.[9]

First, doing so furnishes a substantive precisification of the idea that inquiry consists in a process of reflection aimed at the provision of a theory. Our model cracks open this black box, identifying not only the starting point and the endpoint, but also two stages in the process that takes inquirers from the former to the latter.

Second, it does this while avoiding parochialism. Our model identifies two stages that do not themselves enforce a commitment to anything like the Cartesian project or Deweyan experimentalism. Nor do the stages render either verboten. Rather, our model represents steps by which inquirers of *any* persuasion, and into *any* topic, transition from the opening of a question to the grasp of a theory that resolves it.

Third, a model that highlights data and theorizing is well positioned to account for the depth and breadth of inquiry. Both features were on display in our discussion of dialectic, which noted that inquiry may incorporate, but is not exhausted by, the provision of arguments. This is just what we should expect if inquiry consists in constructing a theory (theorizing) that handles a potentially wide range of considerations (data).

Fourth, our model makes sense of a success condition on theories noted in the Introduction: a theory is successful only when it handles the data, its inputs. It also explains the Holmesian

---

[9] In a pair of rich but neglected essays on philosophical methodology, Collingwood (1933, Ch. VIII) and Castañeda (1980, *passim*) also insist on the importance of data collection and theorizing. However, neither presents them as the two basic stages of inquiry, and their conception of each is markedly different from ours.

injunction against theorizing in the absence of any data. Additionally, the model sheds light on various aspects of particular real-world inquiries, such as the ways in which they can be incomplete (as when they omit data collection, exclude theorizing, or make only partial contributions to either) or deficient (as when they do at least one of these things poorly). In keeping with these remarks, we're not intent on gate-keeping: incomplete or deficient inquiry still counts as inquiry.

Fifth, by distinguishing the roles played by data collection and theorizing, our model makes it possible to examine the basic elements of inquiry and their relations in particular cases. Even though many actual instances of inquiry only imperfectly realize the two stages, attending to them helps to sharpen questions about how, in practice, to get from the opening of inquiry to its closing. In other words, modeling inquiry in terms of these two stages is not just theoretically interesting but also practically important. For it enables us to scrutinize what it takes to do *each*, and to do each *well*—an important first step toward improving our efforts at inquiry.

To appreciate how distinguishing the two stages does this, notice that it helps us to pinpoint various methodological commitments, which can then be subject to proper assessment. In philosophy, the longstanding debate between metaethical realists and their expressivist rivals can serve as an illustration.

Realist views affirm the existence of a moral reality that is objective, in the sense that it is not of our own devising in the way that (say) the norms of etiquette are. And they hold that moral thought and discourse sometimes genuinely represent this reality. Expressivists typically deny at least one of these claims. Divergent methodological commitments lie at the heart of this disagreement.

Each view makes assumptions about *data*. Realists typically assume that the metaethical data consist of commonsensical or intuitive claims, some of which appear to imply that there is an objective moral reality that moral thought and discourse represent. Expressivists ordinarily assume that the data primarily concern moral language and moral judgment, emphasizing connections among moral discourse, moral thought, and action.

Likewise, each view makes assumptions about *theorizing*. Realists typically hold that a satisfactory theory ought to endorse the implications of common sense or intuition, abandoning them only when challenges to them cannot be adequately met. Expressivists, for their part, typically assume that a satisfactory theory must be parsimonious, not positing any features that are explanatorily dispensable. They add that a satisfactory theory must be sufficiently continuous with what the natural sciences tell us about the world.

These assumptions are not idle; rather, they have fueled much of the debate between realists and expressivists. Adherents of each view draw very different conclusions largely because of their divergent assumptions about data and theorizing. Realists typically conclude that since their theory endorses the implications of common sense or intuition, and has adequately met the relevant challenges, theirs is the view to beat.[10] Expressivists ordinarily conclude that their view alone handles the data about language

---

[10] See, e.g., Nagel (1986, 143–4), McNaughton (1988, 40–1), Brink (1989, Chs. 1–2), Shafer-Landau (2003, Chs. 1–3), and Enoch (2019, 30–3). While not a realist, Harman (2010) promotes a general conservative methodology that squares with the approach favored by these authors.

and judgment in a way that is sufficiently parsimonious and naturalistic, and therefore earns top marks.[11]

Realists and expressivists often fail to make their commitments regarding data and theorizing explicit, and rarely pause to defend them even when they do. Nor is metaethical debate unique in these respects. To identify just two other examples: the debates between realists and nominalists about universals exhibit a similar dynamic, as do those between defenders and critics of the a priori.[12] The former members of each pair frequently begin with an unarticulated commitment to intuitive data and conservative theorizing, while the second regularly assume that it is safe to prefer a theory that is both parsimonious and naturalistic.

Here we take a different approach. Though our remarks thus far have ranged over all kinds of theoretical inquiry, our aim will be to explicitly state and defend a series of methodological commitments about data and theorizing that, in our view, anyone engaged in philosophical inquiry should accept. We'll articulate these commitments over the coming chapters, offering detailed accounts of the nature of data, procedures for data collection, and the process of theorizing. Along the way we'll indicate how our commitments relate to those of pragmatists, conceptual analysts, experimental philosophers, proponents of reflective equilibrium, fans of modeling, and various others. As will become evident, the commitments we advocate are both substantive and diverge in important ways from those that frequently guide philosophical inquiry. At the same time, we do not discard current practices, but rather gather

---

[11] See, e.g., Blackburn (1993, Part II and 1998, Ch. 3) and Gibbard (2003, Chs. 1–2). *Cp.* Ayer (1936, Ch. 6), who adds a commitment to a now-defunct theory of meaning.

[12] See, e.g., Bealer (1993), Dorr (2007), and Carmichael (2010) on universals, and the exchange between Bonjour (2014) and Devitt (2014) on the a priori.

and organize the bits of methodological wisdom they contain. We'll argue that when philosophers adhere to the commitments we present, the theories they develop are poised to realize the epistemic goals of inquiry. It is to these goals that we now turn.

2. The Goals of Inquiry

Aristotle's *Posterior Analytics* is naturally read as proposing that an ultimate proper goal of theoretical inquiry is the acquisition of *epistêmê*—where this is not mere ordinary knowledge, but a special kind of epistemic ideal that many commentators designate with the term 'understanding.'[13] We endorse a version of this proposal.

Let us begin by explaining what we have in mind by an 'ultimate proper goal' of theoretical inquiry. We allow that the set of possible goals of inquiry is open-ended. It may include mere psychological states, such as what Charles Sanders Peirce referred to as "the fixation of belief,"[14] or accomplishments such as happiness or salvation. No doubt it includes positive epistemic states such as justified belief, knowledge, and understanding.

We take some such goals to be 'proper' because, while there might be no such thing as *the* correct goal of a given activity, some goals are more fitting to an activity than others. This is particularly true of goals whose realization constitutes a type of success or excellence at that activity. As an example, consider investigative journalism, which has a variety of proper goals that include uncovering official malfeasance, identifying social inequalities, or

---

[13] See, e.g., Burnyeat (1981, 102–3 and 112), who describes Aristotle's proposal as the "vision of Plato's *Republic* transferred to the individual sciences."

[14] Peirce (1877, §4).

exposing reparable inefficiencies within an organization. Such journalism is done well when it succeeds at these aims. By contrast, humiliating personal enemies or publicizing boring arcana are improper journalistic goals, for their achievement does not imply a job well done. When the activity is theoretical inquiry, undertaken for its own sake, success or excellence does not consist in scoring points, winning accolades, or cultivating good habits. Rather, it resides in making an intellectual improvement, in the form of an epistemic or alethic achievement. In our view, this is the *têlos* of inquiry; it is what inquiry is *for*.

We call a proper goal 'ultimate' just in case its realization is sufficient to resolve inquiry in a fully successful way, reaching the point at which, to put it simply, there is no more work to be done. The notion of the fully successful resolution of inquiry is not purely descriptive, but has a normative component as well, which admits of various explications. According to one plausible explication, such resolution renders it appropriate for the inquirers' curiosity, wonder, or puzzlement to be relieved thereby.[15] Though we wish to remain neutral between this and other candidate explications, we affirm that any plausible explication will demand not merely partial and piecemeal insights, but rather comprehensive and systematic illumination.

Not all proper goals are ultimate. For example, to be conscientious at each stage of one's investigation may be a proper goal of inquiry, since it represents a type of success or excellence at this activity. But conscientious inquiry may produce an erroneous picture of its target, and so does not qualify as an ultimate proper

---

[15] Pritchard (2016, 34). We are wary of non-normative explications, such as those that emphasize the first-personal experience of a "Eureka moment" (*cp.* Trout 2002).

goal. Even if inquirers were to correct these errors, their job would hardly be over, since true belief can be unjustified and still leave open many of the most significant questions about the target of inquiry.[16] Were inquirers to attain a true belief that was also justified, there would remain more work to do. For such a belief, or even a set of such beliefs, does not by itself guarantee the sort of comprehensive and systematic illumination required to resolve inquiry in a fully successful way (hereafter, simply 'resolve' inquiry).

The point becomes clear when we consider that even knowledge fails to secure the guarantee. For such an achievement might be gained on the basis of testimony or rote memorization, rendering it compatible with obliviousness to the considerations that support or explain the answer. Ordinary knowledge is also congruent with messes or tangles, as when new information does not resolve cognitive dissonance, but instead *induces* it. In either case, inquiry isn't finished.

Importantly, knowledge also may incorporate confusion and muddle of various sorts. A grade-school class might know that π is

---

[16] We thereby reject Dubois' (1898, 16) contention that the aim of inquiry is "simple truth," a position echoed by a number of contemporary philosophers (*cp.* Niiniluoto 1984; Haack 1993; Anderson 1995). Other critics of this view include Kitcher (1990 and 1993), Bright (2017), and Nado (2019), as well as proponents of the simple knowledge view discussed in the text (e.g., Williamson 2000; Bird 2007; Whitcomb 2010; Mirrachi 2015; Beebee 2018; Sosa 2019; Kelp 2021). There are a clutch of alternative conceptions of the aim of philosophical inquiry in particular. For example, Gutting (2009 and 2016, 325) privileges "knowledge of distinctions and of the strengths and weaknesses of various pictures and their theoretical formulations"; Nozick (1981, 21–2) and Wilson (2014, 145–50 and 2017, 92) celebrate the amassing of theoretical options; Deleuze and Guattari (1996, 5) hold that "the object of philosophy is to create concepts that are always new"; Adorno (1993, 102) views philosophy as "an effort to express things one cannot speak about"; and many philosophers insist that securing agreement is part of philosophy's aim. The considerations that follow apply *mutatis mutandis* to these proposals; we raise doubts about the need for agreement in §3.

the ratio of a circle's circumference to its diameter, while being confused about the concept RATIO. For instance, they could mistakenly conceive of a ratio as (say) something whose decimal expansion must terminate or repeat. In such a case, they would know the answer to a key question about π—"Is π the ratio of a circle's circumference to its diameter?"—without having reached the point at which further inquiry regarding this question would be otiose. Philosophers who know the answer to a central question in the philosophy of race—"Is race a social construct?"—could be in a similar position. Whatever the correct answer, if they are confused about the concept SOCIAL CONSTRUCT, mistakenly conceiving of it as (say) requiring explicit group agreement, their inquiry regarding this central question isn't closed. For there would be cracks in their grasp of the answer that make it unfitting to call it a day; their knowledge of the answer would not appropriately relieve their curiosity.

The core problem is not solved by piling on additional knowledge—for instance, knowledge of what a ratio or social construct is. For any such knowledge could itself incorporate conceptual confusions that render it defective. It follows that thinkers could have wide-ranging knowledge of the answers to the questions prompting their inquiries, but fail to resolve them. While ordinary knowledge of an answer is a proper goal, it falls short of resolving inquiry, and thus being an ultimate proper goal in our sense.[17]

Inquirers don't *need* to pursue such a goal—proper goals are good, too! Still, the differences between the various goals are

---

[17] This line of reasoning entails the rejection of two norms of inquiry proposed by Friedman (2017 and 2019): the "Don't Believe and Inquire" norm and the "Ignorance Norm," according to which one ought not inquire, if one believes or knows, respectively. *Cp.* Archer (2018, §5).

significant, given three purposes that they serve. Such goals furnish reference points for selecting promising questions to pursue in the course of inquiry, as these questions are ones whose resolution will secure those goals; they provide a yardstick for assessing the merits of theories, which are criticizable to the extent that they do not deliver the goods that such goals specify; they also function as the basis for evaluating the methods used to construct those theories.

For example, a method of theorizing might yield true belief and nothing more. But such a method is inadequate if what we are seeking is an ultimate proper goal of inquiry. If the outputs of theorizing should be assessed, as we believe, by reference to whether they enable realization of such a goal, then when evaluating extant philosophical methods (Chapter 4) and developing and defending our own preferred method (Chapters 5 and 6), we'll want to have squarely in view the question of whether they manage this feat. It will be helpful to have a term for a method that does. Accordingly, we propose to call a method 'sound' just in case satisfaction of its criteria thereby positions inquirers to achieve an ultimate proper goal of inquiry. If a method is sound, then (ceteris paribus) inquirers have strong reason to endorse its outputs.

Our own view, as noted, is that theoretical understanding is an ultimate proper goal. We turn now to articulating the central characteristics of the relevant type of understanding, before arguing for its appeal as such a goal.

3. Theoretical Understanding

We can reveal the central characteristics of understanding by considering some of the properties that a theory of a given domain must possess in order to provide understanding of that domain to thinkers who fully grasp the theory.[18] (We're working under the assumption that a domain, or subject matter, consists of objects, properties, and the like, and is individuated by a set of questions. We'll speak sometimes of understanding a domain and at other times of understanding with respect to a set of questions.[19]) Given the number of controversies surrounding the nature of understanding, we wish to make clear which features we take this epistemic achievement to exemplify. Theoretical understanding, as we'll construe it, is the state that agents possess just when they fully grasp a theory with the following six properties.

First, the theory possesses a high degree of *accuracy*, since largely inaccurate theories will fail to dispel confusion (a characteristic of misunderstanding).

Second, the theory is *reason-based*, in the sense that it is positively supported by considerations, beyond mere coherence, that speak in favor of its accuracy. For in the absence of such support, signing on to the theory would be arbitrary or haphazard (again, a characteristic of misunderstanding).

Third, the theory is *robust*, answering a multitude of questions about the most important features of the domain under investigation. A theory that neglects or dodges such questions leaves out just what's needed to yield comprehension.

---

[18] We speak of 'fully' grasping not to imply perfection in every respect, but rather to draw a contrast with a merely partial grasp.

[19] Like inquiries, domains can overlap with, nest in, and compose one another. They can also be cognized independently of fully successful inquiry into the set of questions that individuate them. While there is much more to say about subject matters, to our knowledge nothing in our project hangs on how they are further theorized.

Fourth, the theory is *illuminating*, in that its answers must at least sometimes be not just general but also genuinely explanatory, going beyond a mere description of those features to explain why each exists or is instantiated.

Fifth, the theory is *orderly*, not simply offering such feature-specific explanations but also affording a broader view of the domain by revealing how those (and other) features, as well as the proposed explanations, gel or hang together—for example, by exposing basic relations or systematic connections among them. Such a theory avoids miscellany, the paradigm of which is a mere list, which says nothing about how, if at all, its various items are ordered or organized.

Sixth, the theory is *coherent*, not only internally but also externally, fitting well with a wide range of understanding-providing theories of other domains. A theory of one subject matter that massively conflicts with a coherent cluster of accurate, reason-based, robust, illuminating, and orderly theories of other domains does not further comprehension but muddles it (yet another characteristic of misunderstanding).

Although all six properties contribute to theoretical understanding, they do so in different ways. The latter two, unlike the former four, only conditionally make such contributions. The orderliness and coherence of a theory contribute to its ability to supply understanding only if the theory possesses the other four features to at least some extent. In this way, these first four are fundamental to understanding in a way that the final pair are not.

When inquirers fully grasp theories with these six properties, the targets of their theories *make sense* to them. This is theoretical understanding, which on our construal is distinctive. It isn't reducible to true or justified belief (or to their conjunction). Nor is it equivalent to ordinary knowledge. For, as explained above, the function of such understanding is, *inter alia*, to illuminate, in a robust, orderly, coherent fashion, the portion of reality under investigation, and not simply to state a series of known truths or justified beliefs about it. Theoretical understanding is not epistemic perfection. Still, it is a type of epistemic excellence. And it is a prime candidate for an ultimate proper goal of inquiry.[20]

To appreciate this last claim, notice that it handles the points that challenged other aspirants. Theoretical understanding could not be unwarranted or narrowly circumscribed, and is neither dissonance-inducing nor compatible with obliviousness. It also excludes the sorts of confusion and muddle illustrated by our examples involving RATIO and SOCIAL CONSTRUCT. That it ticks these boxes is unsurprising. After all, were inquirers to fully grasp a theory that is accurate, reason-based, robust, illuminating, orderly, and coherent, their grip on the inquiry's subject matter would be both comprehensive and systematic. They would possess accurate replies to a wide range of central questions about that subject matter, including explanatory ones. Their answers would fit together, and enjoy rational support. Moreover, the inquirers would grasp those answers without conceptual confusion. In such a case, their curiosity, wonder, or puzzlement would be appropriately relieved.

---

[20] Recent epistemology has witnessed a surge of arguments for the value of understanding over all other epistemic achievements: see, among others, Elgin (1996, 122ff.), Zagzebski (2001), Kvanvig (2003, Ch. 8), and Pritchard (2010, Ch. 4 and 2016, §§2–4); *cp*. Dellsén (2016). We will not rely on these arguments here, though they may offer independent support for our assessment. See Bengson (2017) for a fuller treatment of understanding.

This line of reasoning may be buttressed by observing that a wide range of more demanding epistemic or alethic achievements, such as absolute certainty, are unnecessary to obtain such relief. Such achievements might be desirable in various respects. But even were inquirers to fall short of them, that wouldn't by itself impugn the success of their inquiries, so long as they terminated in full grasp of theories with the six properties we've identified.

Our arguments in the previous section also cast doubt on the claim that there is a less demanding epistemic or alethic achievement that is an ultimate proper goal. Though we certainly haven't offered a decisive refutation of this claim, we have made the case that ordinary knowledge fails to resolve inquiry. (We've also given reason to believe that less demanding states, such as true belief, justified belief, and justified true belief, fail as well.) So any ultimate proper goal of inquiry distinct from understanding would have to be stronger than knowledge, but at least as strong as understanding. We're not sure what that might be—especially given our contention above that anything stronger than understanding is unnecessary. In any event, there is an important sense in which this doesn't matter. For we've provided reason to think that understanding is the minimal epistemic achievement needed to resolve inquiry. And that's all our arguments in this book require.

(In order to streamline our discussion, we'll often write as if theoretical understanding is inquiry's ultimate proper goal. This should not be read as implying that there aren't any other ultimate proper goals, but rather as shorthand for the claim that understanding suffices to resolve inquiry, and no weaker epistemic or alethic achievement can do so.)

The special role we are attributing to understanding can be further motivated through a contrast with the popular idea—recently defended by David Chalmers—that inquirers ought to aim for *collective convergence on the truth*. Chalmers contends that other possible goals, including understanding, involve "something of a lowering of our sights."[21] While we endorse the emphasis Chalmers places on collectivity, we reject a narrow focus on truth and agreement. In addition to concerns raised earlier (in §2) about the relation between true belief and inquiry's ultimate proper goal, we'd like to register two further worries about Chalmers' proposal. Both rely on drawing attention to the six understanding-enabling properties we've identified.

First, collective convergence on the truth appears to be insufficient to resolve inquiry. For such convergence would not ensure possession, let alone full grasp, of a theory that possesses all six of the features described above. But in the absence of this, we could hardly deem inquiry complete. There would remain more work to do.

Second, convergence is unnecessary, and this is so whether it consists in full or partial agreement. Suppose we discovered that human psychology (with its familiar penchant for generating disagreement) rendered convergence unattainable. Still, were inquirers fortunate enough to develop a theory that realized all six features associated with theoretical understanding, fully grasping that theory would suffice to resolve inquiry, even if agreement remained elusive.

We conclude that Chalmers' charge gets things backwards. Convergence on the truth closes inquiry only when it is accompanied by understanding. In this sense, it is aiming for mere agreement on the truth that would be "something of a lowering of our sights."

---

[21] Chalmers (2014, 14). By 'the truth,' Chalmers is referring to the proposition that correctly answers the question under investigation.

That said, as anticipated above, we support Chalmers' suggestion that inquiry's aspiration be couched in collective terms. Specifically, we affirm that an ultimate proper goal of inquiry must be both capable of being pursued collectively and available to collectives. Theoretical understanding meets these conditions. It can be realized through collective efforts at data-gathering, theorizing, or both. And the understanding that results from those efforts is in principle widely available.

We've been focusing on the importance of understanding to theoretical inquiry. It may be that science, philosophy, and neighboring disciplines incorporate other, non-epistemic projects as well—e.g., those that are aesthetic, felicific, salvific, pragmatic, or political. No doubt philosophers have at times appraised competing views by these (and still other) non-epistemic standards. Our intention is not to question the legitimacy of such appraisals, or to promote a picture of inquiry on which its sole interest derives from the ambition of attaining theoretical understanding with respect to its central questions. Rather, our contention is simply that such understanding is theoretical inquiry's ultimate proper goal.

This contention should be congenial to those with disparate metaphilosophical commitments. Granted, many pragmatists, intellectual anarchists, and anti-theorists *appear* to decry the sorts of ambitions we have set for inquiry.[22] But their misgivings are

---

[22] All three positions have long been thought to have severe metaphilosophical implications. For two influential works by anti-theorists, see Baier (1985, Ch. 12) and Williams (1985). Dewey (1910 and 1938) is a well-known pragmatist, and Feyerabend (1975) famously avows anarchism. Urmson (1967, 232) also gives voice to the anarchist view in his summary of J. L. Austin's work: "No one knows what a satisfactory solution to such problems as those of free will, truth, and human personality would look like, and it would be baseless dogmatism to lay down in advance any principles for the proper method of solving them."

typically directed at ideas to which we are not committed or that we explicitly reject, such as an exaggerated sense of the importance of simplicity or an inflated notion of objectivity as requiring an "absolute conception of the world." Moreover, our approach coheres with the stated motivations of many pragmatists, anarchists, and anti-theorists. These motivations include the historical and cultural situatedness of reflective thought, and the wide variety of considerations relevant to such reflection. Our contention is also compatible with the position that, in some cases, theoretical understanding with respect to certain questions is not achievable, or at least is not readily available. We agree, for instance, with David Wiggins when he writes that it is often "a matter of prolonged and difficult inquiry gradually to improve currently accepted standards or conceptions" in philosophy.[23]

If our contention is correct, two conclusions follow. First, a method is sound only if it possesses the resources to guide the construction of a theory that positions inquirers to acquire understanding, and (concomitantly) to facilitate evaluations of candidate theories in this light. Second, because methods take data as their input, we must get clear about the nature of data and the means by which they are to be collected. We devote the next two chapters to presenting views about data in philosophy that are consonant with their role in developing a philosophical method poised to yield successful theories. We'll then devote the remainder of the book to assessing extant methods, presenting the method that we prefer, and revealing its many benefits.

---

[23] Wiggins (2001, 82).

2

Philosophical Data

Every theoretical inquiry has its data, considerations that must somehow be handled by a theory that resolves the questions that open inquiry. Across philosophy, examples abound: in philosophy of language, it is a datum that a user of a proper name sometimes knows next to nothing about the person being named; in philosophy of perception, that we sometimes suffer vivid illusions and hallucinations; in philosophy of action, that some attempts are unsuccessful; in metaethics, that moral judgments are intimately connected to action; in aesthetics, that certain works of art are masterpieces; in feminist philosophy, that a range of social practices and institutions promote the interests of men over women; and so on. Our aim in this chapter is to achieve greater clarity about the character and role of such data.

We argue for three main claims. First, philosophical data possess four basic features, which set them apart from other types of considerations. Second, familiar worries about the

Philosophical Methodology: From Data to Theory. John Bengson, Terence Cuneo, and Russ Shafer-Landau, Oxford University Press. © John Bengson, Terence Cuneo, and Russ Shafer-Landau 2022.
DOI: 10.1093/oso/9780192862464.003.0003

"theory-ladenness" of data are not devastating impediments to philosophical theorizing, but can be plausibly addressed. Third, philosophical data should be conceived of not as what inquirers agree upon, constitutive features of reality, claims about how things appear, or descriptions of how language is commonly used. Rather, as we argue in the next chapter, they are what inquirers have good epistemic reason to take to be genuine features of reality itself.

1. Four Features of Data

Data possess four basic features, which will play important roles in the ensuing discussion.

First, data are *starting points* for theoretical reflection on a domain in the sense that they are inputs, not outputs, of such theorizing. In this respect, the data in a given domain bear an asymmetrical relation to theorizing about that domain.

Second, data are *inquiry-constraining* with respect to a domain in the sense that they comprise a data set that functions to anchor a given theoretical inquiry to its subject matter, the domain that such inquiry is about. By saying that they 'anchor' inquiry to its subject matter, we intend to convey two points. First, the data set operates as the principal means by which theorists access that subject matter. Second, investigations of a domain that entirely ignore the data set are likely, or at least significantly likelier, to be off-track with respect to that domain, and consequently to fail to resolve inquiry. Although ignoring one datum might not send inquirers on a wild goose chase, those who wholly ignore the data pool are much likelier to lose the trail.

Third, data are *collected*. While not itself a controversial claim, there is controversy regarding the collection of data, especially over appropriate procedures with which to pursue such a task. Among the candidate procedures for data collection are those that utilize such sources as perception, intuition, introspection, common sense, linguistic judgment, imagination, and inference. There is room to be more or less restrictive on this matter; some philosophers recognize only one of these as part of an appropriate procedure for data collection, while others embrace a pluralism that allows for a variety of such procedures. (We devote part of the next chapter to exploring these matters, and to defending a version of pluralism.)

Fourth, data are *neutral*. There are two respects in which this is so. The first is that data are non-factive, at least insofar as a particular datum might be mistaken. Even if a pool of data is on the whole correct, there are sometimes errors in the data. A data set may be noisy or include outliers that are incompatible with other elements of that set. Data are inputs to theorizing, and such inputs may not survive scrutiny when all is said and done.[1]

The second respect in which data are neutral is that they function as common currency among theorists. There are several

---

[1] Our reasons for viewing data as non-factive do not rule out that there is a distinct use of the term 'data' according to which the sentence 'data are always correct' is true. Perhaps this is how to interpret Williamson's tendency to use 'data' interchangeably with 'evidence' (see, e.g., 2007, 5 and 191), which he maintains is factive (2000; but see 2018, 78–80, where this tendency is accompanied by recognition that "the data usually contain some random errors"). That, however, is not our use. As we've indicated in the text, data are inputs to a method of theorizing—starting points whose epistemic status is left unspecified by our statement of its four basic features—whereas evidence is a positive epistemic status that is not restricted to starting points. We comment on the relation between data and evidence in Chapter 3 (§1).

dimensions to this. For one thing, data are the sorts of things that merit attention from theorists of diverse persuasions, even if some of those theorists ultimately reject them. For another thing, data underdetermine theory: this fact enables inquirers of various stripes to converge on a pool of data while at the same time exploring different ways of handling them.[2] Data are also frequently open-ended: theorists may differ about how to interpret, refine, or clarify the relevant data, which can be vague or underspecified, and so in need of precisification. In these ways, data are ecumenical, uniting theorists engaged in a common enterprise.

What we've been referring to as 'the four basic features of data' are, in effect, data about data. That is, in the context of theoretical inquiry into the nature of data, these features are neutral, inquiry-constraining starting points that we have collected through reflection on what data are like and the role they are meant to play. Like many other data, some of these four features admit of further precisifications, which may be controversial; we'll not wade into such debates here.

Such precisifications must be distinguished from competing *theories* of data, to be canvassed momentarily. Before considering them, we must confront what is likely the most serious challenge to the very notion of data, namely, that all putative data are inevitably theory-laden, and so unfit to play a guiding role in inquiry. As we'll now argue, thinking of the four basic features as data about data has an important upshot: it enables us to pinpoint what exactly is at issue in these familiar worries and to identify plausible ways of defusing them.

---

[2] *Cp.* Mill (1867/1900, 328), Duhem (1914), and Quine (1970, 179), who noted that "Theory can still vary though all possible observations be fixed." We return to the underdetermination of theory by data at several points below.

2. The Theory-Ladenness of Data

Philosophers have long worried that data are not neutral starting points but instead are theory-laden.[3] In that case, the data would be suspect, unduly stacking the deck in favor of a particular theory; they couldn't play the role that data are supposed to play. Our characterization of the basic features of data in the previous section helps to sharpen the challenge. There we noted that data are starting points, bearing an asymmetrical relation to theorizing. They are also neutral, being not only non-factive but also common currency among theorists. Each of these features is threatened if the data have already sided with a given theory. That is why the charge that data are theory-laden raises a worry about the very possibility of data.

There are several things to say in reply to this concern. Note, first, that the idea of the data's being 'theory-laden' could mean that the data are *infused* with a theory, or instead *partial* to a theory. These are importantly different. But neither, we think, is inherently troublesome. The latter needn't itself be in conflict with our description of data as neutral starting points, as there are familiar cases of unproblematic partiality. That physical data are partial to relativistic theory, for example, does not imply that the data are somehow rigged when theorizing about the subject matter of physics, or that they are the results of—rather than inputs to—such theorizing. Likewise, infusion needn't be problematic, provided that the data are infused with commitments from well-supported theories in domains beyond the one under investigation. Consider,

---

[3] See, e.g., Hanson (1958, Ch. 1) and Kuhn (1962, §X). Worries about theory-ladenness are not restricted to scientific data but arise for philosophical ones as well; see, e.g., Wittgenstein (1953, II.vi) and MacIntyre (1988, 333).

for example, a datum from the metaphysics of personal identity: we have mammalian bodies possessed of a four-chambered heart, a brain composed of approximately a hundred billion neurons, and other organs. Such a datum is saturated by commitments from biology and anatomy, but this does not undermine the datum's status as a neutral arbiter between competing views of personal identity. Data are collected, and a good source of data collection may make use of well-supported theories in other domains in order to get the job done.

To be clear, we are not denying that pernicious forms of theory-ladenness are possible. However, and as a second reply to the concern, it is often relatively easy to identify such cases, and to nullify or correct for the adverse influences. For example, confronted with the worry that a given consideration is biased, proponents of very different theories might undertake the same experiments. If they were to arrive at similar results, then that would tell against any such bias. Some cases might of course slip by unnoticed. Perhaps certain considerations are infused by the very theory that they are claimed to support, or by theoretical commitments that are arbitrary or unjustified. Such considerations would indeed fail to qualify as data (inputs to theorizing), precisely because of that infusion. But absent reason to believe otherwise, it would be overly pessimistic to think that such pernicious influences are pervasive or incorrigible.

While we take the foregoing points to answer general worries about theory-ladenness, we also note a third reply. Suppose, as seems plausible, that some data are partial to certain aesthetic theories over others. For instance, suppose the intimate connection between aesthetic judgments and personal preferences is more congenial to subjectivist views than objectivist ones, whereas the existence of art experts and the longstanding recognition of certain works as masterpieces tilts in the opposite direction. Objectivists should and do acknowledge the former datum; subjectivists should and do concede the second. But so long as the methods that work with the data are themselves neutral among competing aesthetic theories, allowing for the rejection of various data under certain conditions, then all is above board.

And that is because, fourth, methods should not regard the data they work with as sacrosanct, in need of preservation come what may. As indicated above, the data are fallible starting points for theorizing. This activity may determine, when all is said and done, that some of the data are mere appearances that do not survive critical scrutiny. That is the verdict issued, for instance, by eliminative materialists—those who deny the existence of anything mental—with regard to the datum that people have beliefs. Alternatively, theorists may accept the data while rejecting the contention that these data are partial to one theory rather than another. Many theists, for instance, accept the datum that bad things happen to good people, while denying that this recommends atheism over their preferred view.

We have provided four reasons not to be worried by the charge that data are bound to be illicitly theory-laden, unable to function as neutral starting points for theorizing. Importantly, the neutrality of data does not imply the type of consensus demanded by the thesis, discussed by Timothy Williamson, that we label 'Strict Neutrality.'[4] This thesis invokes an especially stringent understanding of

---

[4] See Williamson (2007, 209ff.), who uses the label 'Evidence Neutrality' and objects to this thesis in the context of a critique of the tendency to "psychologize" data. We also reject this tendency; see §3.3 below.

neutrality according to which a claim φ is neutral only if φ's truth-value is "in principle uncontentiously decidable," in the sense that one is "able to persuade all comers, however strange their views," of φ's truth-value.[5] By contrast, our notion of neutrality has no such entailment. Strict Neutrality certainly does not follow from the idea that data are non-factive. Nor is it implied by the claim that data function as common currency. Compare: dollars are common currency in North America, even though some merchants, restricting their transactions to those involving gold, axes, skins, or fish, may refuse under any conditions to recognize dollars as legitimate.

Illustration 2　Nordic and Russian traders bartering.

Consequently, our notion of common currency helps to preserve the role of a data set as "a neutral arbiter between rival theories."[6]

However, the main problem with Strict Neutrality as applied to data is not that it is overkill, but that it dooms inquiry. Given philosophers' ability to come up with reasons to doubt even the

---

[5]　Williamson (2007, 210–14).

[6]　Williamson (2007, 210). This is a role that Williamson himself applauds, although it is not clear how well his view preserves it, since his view identifies data with evidence and evidence with knowledge (a factive state), as discussed in note 1.

most obvious claims, few if any considerations could garner the type of consensus demanded by this thesis. Strict Neutrality shrinks the data to a vanishing point, thereby undermining the very possibility of inquiry.

3. Three Theories of Data

Philosophers have rarely engaged in direct and explicit theorizing about philosophical data—about what data are or why a datum qualifies as such. Still, it is possible to identify several very different theories of data. This section presents and critically assesses three of the most salient candidates, evaluating each in light of the extent to which it handles the four basic features of data (the data about data). These theories treat data as what inquirers agree on, constitutive features of reality, or claims about language or appearances. We'll argue that all three of these views are problematic. The next chapter presents and defends an alternative theory, which focuses on data's epistemic profile.

Three points before we undertake the examination. First, our formulation of each of the theories will include a clause intended to ensure that data are pre-theoretical, entering the scene at stage one of inquiry rather than at stage two, which concerns theorizing. The clause says that the data regarding a domain D do not belong to any 'well-formed' theory of D. That is, they are not members of a set of claims regarding D that satisfy a sound method's criteria at the second stage of inquiry into D.[7] In this sense,

---

[7] This definition of a well-formed theory builds upon our characterization of a theory in Chapter 1 (§1). The fact that bad theories may simply repeat the data motivates our focus on theories that are well-formed (in our technical sense), which handle the data.

data are pre-theoretical. While they may be partial to one or another theory about a given topic (as noted in the previous section), they do not close inquiry regarding that topic.

Second, our assessment of theories of data will employ a fairly open-ended understanding of what it takes to *handle* a datum. If a theory of data makes sense of or explains various data about data, or if it delivers the result that they hold, then it handles them. However, if a theory is incompatible with some of these data or provides no assurance that they hold, then the theory does *not* handle them—at least provided that no justification for their rejection is forthcoming. (We'll offer a detailed account of what it is to handle a datum in Chapter 5, but for now this gloss will suffice.)

Third, our evaluation will focus primarily on whether the theories handle just three of the four data about data: that they are starting points, neutral, and inquiry-constraining. All of the theories we'll canvass are compatible with the remaining datum—that data are collected—though none of them does much to make sense of why data are collected or to explain such collection. As discussed in the next chapter, we believe that our preferred view of data does well on both counts.

3.1 The sociological theory

According to what we call the 'sociological theory' of data, φ is a datum regarding domain D if, only if, and because φ is about D, does not belong to any well-formed theory of D, and is a claim that inquirers provisionally agree upon as central to D.[8]

---

[8] See, e.g., Heney (2016, 26). *Cp.* the "pragmatic view" discussed by Williamson (2007, 238).

The sociological theory smoothly handles the observation that data are starting points. It may also account for the neutrality of data, since a claim that is mistaken might satisfy its conditions (per non-factivity), and what satisfies its conditions could be a fitting object of attention for inquirers of disparate theoretical persuasions (per common currency).

However, the sociological theory does not square with the idea that data are inquiry-constraining. For the view does not provide assurance that the data set will anchor inquiry to its subject matter by both facilitating access to that subject matter and ensuring that neglect of that set is more likely to render theorizing off-track. The reason is straightforward: inquirers' agreements do not determine what belongs to a domain. Suppose a community of philosophers seeking to answer the central questions about justice were provisionally to agree that justice is identical to an even number. (Derek Parfit reports that some Pythagoreans are reputed to have shared this view.[9]) It would not follow that this claim is a datum about justice—contrary to what the sociological theory implies.

It is fair to say that if the data were in fact to relate inquiry to its subject matter, then this would be an accident were the sociological theory true.[10] While the coincidence would be a happy one,

[9] Parfit (2011, 324).

[10] Presumably this counterfactual would be false just if the relevant worlds of assessment include idealized agents, such as those who are superlative truth-trackers, or instead those who enjoy a high degree of positive epistemic standing. But the first idealization threatens to render the sociological view a version of the metaphysical theory (discussed in §3.2), while the second threatens to make it a species of the epistemic theory (developed in the next chapter). In any case, such an idealization is in tension with the spirit, if not the letter, of the sociological theory, which aspires to recognize the highly *non*-ideal conditions within which humans inquire.

it would be lucky nevertheless. That, however, is at odds with data's status as inquiry-constraining.

This conflict is only to be expected. For the sociological theory allows that data (to borrow a memorable phrase from Richard Rorty) closely track "what our peers let us get away with."[11] With perhaps the singular exception of inquiry regarding what our peers let us get away with, this falls far short of anchoring inquiry to its subject matter in the relevant sense.

3.2 The metaphysical theory

What we call the 'metaphysical theory' of data holds that φ is a datum regarding domain D if, only if, and because φ is about D, does not belong to any well-formed theory of D, and is a claim that (expresses a fact that) is genuinely constitutive of D.

The metaphysical theory implies that a domain owes its existence and character to the data about that domain. To illustrate, recall the data from aesthetics mentioned above: there is an intimate connection between aesthetic judgments and personal preferences, there are art experts, and certain works of art are masterpieces. According to the metaphysical theory of data, these data partly constitute the aesthetic domain. That domain exists and is what it is in virtue of them.

While this view of data has not to our knowledge been explicitly articulated, it seems to present an attractive alternative to the sociological theory. The metaphysical view makes sense of the ideas that data are starting points and inquiry-constraining, for it provides a pre-theoretical metaphysical anchor ensuring that the data regarding a given domain allow inquirers to access that domain.

---

[11] Rorty (1979, 176).

However, this view belies the fact that some data are *not* constitutive of the domain about which they are data. Take, for example, the observations that one agent's belief may contradict another's, that beliefs and desires can have incompatible contents, or that some beliefs are less interesting than others. These observations do not identify very good candidates for what is constitutive of belief, although they do locate data about belief, considerations that any theory of belief should be able to handle.

Perhaps more flagrantly, the metaphysical theory implies that the data are never mistaken, which directly contradicts the non-factivity—and so the neutrality—of data. Proponents of the metaphysical theory might seek to reinstate neutrality by introducing a prior phase of data collection that aspires to identify what inquirers have reason to believe to be constitutive of a given domain, where such a reason is non-factive. But as will become clear, this is effectively to abandon the metaphysical theory in favor of the epistemic one that we favor.

3.3 The psycho-linguistic theory

According to what we dub the 'psycho-linguistic theory,' φ is a datum regarding domain D if, only if, and because φ is about D, does not belong to any well-formed theory of D, and belongs to a domain-specific class of psychological or linguistic claims.

In the paradigm cases, these are claims about how things appear, what intuitions we have, what we pre-theoretically believe, or how language is aptly or commonly used. For example, within metaphysics, it is widely accepted that colors often *appear* to be objective and that it is apt to *say*—perhaps when discussing the contents of a box of crayons—"It's a fact that this crayon is red while that other one is orange."

The psycho-linguistic theory holds that *all* data are like this.[12] To illustrate, consider a more precisely formulated version of a putative aesthetic datum mentioned above, namely:

Preferences: Aesthetic judgments are intimately connected to personal preferences.

This formulation makes no explicit reference to how things appear, what intuitions we have, what we pre-theoretically believe, or how language is used; it tells us about a putative fact regarding the intimate connection between aesthetic judgments and personal preferences. Now consider:

Preferences$_\psi$: It *appears* that aesthetic judgments are intimately connected to personal preferences.

Preferences$_\lambda$: It is apt to *say* (directly or by implication), "Aesthetic judgments are intimately connected to personal preferences."

These data make explicit reference to how things appear or how language is used; they do not concern a putative fact regarding the intimate connection between aesthetic judgments and personal preferences. The psycho-linguistic theory understands data to take the form not of Preferences but Preferences$_\psi$ or Preferences$_\lambda$.

The psycho-linguistic theory can readily handle the observation that data are starting points. However, the principal motivation for this view's restriction of the data to psychological or

---

[12] Some philosophers endorse restricted versions of the psycho-linguistic theory, which allow that some data are not psycho-linguistic, even though the central ones are (see Castañeda 1980, 32 and 47; for domain-specific examples, see Weatherson 2003, 27 and Finlay 2014, 121). The objections we'll press below apply equally to restricted versions.

linguistic claims is to satisfy neutrality. If the data consist in such claims, then those data might function as common currency among otherwise rival theorists, whose disputes do not impede their agreement about such things as appearances or linguistic usage. Those data may also be mistaken, failing to depict genuine features of the domain under investigation, which are not entirely psychological or linguistic.[13]

This virtue of the psycho-linguistic theory is, however, intertwined with what is perhaps its greatest vice, concerning the relation between data and their associated subject matters. The problem can be brought out by the following dilemma.

Suppose, first, that the psycho-linguistic theory holds that the data themselves are the relevant subject matters. This would imply that all of the subject matters of philosophy consist entirely in psychological or linguistic phenomena. Of course, that is precisely how psychological and linguistic idealists conceive of things. But idealism of any stripe is a substantive *theory* (which itself results from theorizing undertaken to handle the data), and a highly revisionary one to boot. Such a theory would demand that we significantly alter our conception of philosophy's subject matter, from something that is often neither psychological nor linguistic to something that is always one or the other. When considering the relation between philosophical data and their associated subject matters, we should refrain from presupposing global idealism.

Turn, then, to the second horn of the dilemma. Suppose that the psycho-linguistic theory were to deny that the data are the subject

---

[13] Many proponents of the psycho-linguistic view seem to be motivated by a commitment to something stronger than neutrality, captured by the thesis of Strict Neutrality. This would render their view vulnerable to the problems of overkill and shrinking data described in §2.

matters of philosophy, holding instead that the data bear some other relation to what it is that philosophical theorizing is about. This is compatible with the data being neutral, and thus accords with the main motivation for the psycho-linguistic approach. It also sidesteps the objection raised just above, since it avoids the implausible consequence that philosophy is invariably about psychological or linguistic phenomena. Indeed, it acknowledges that philosophy's subject matter often consists in objects, properties, facts, and other entities that are not psycho-linguistic. However, we see three problems with this way of proceeding.

First, this approach posits a sizeable gap between the data and the subject matters of philosophy. To be clear, this gap is about two disparate types of content—psycho-linguistic content and the subject matters of philosophy, respectively. Given that data are inquiry-constraining, this gap cannot remain impassable; it needs to be bridged, so as to assure that those inquiries are anchored to their subject matters, which are not exclusively psycho-linguistic. Here we find a deep tension within the psycho-linguistic theory. On one hand, it requires there to be a substantial gap between the contents of the data and the contents of a given philosophical domain, both to fulfill its aspiration to secure neutrality and to avoid omitting huge swaths of that domain. On the other hand, it also needs there *not* to be a substantial gap, in order to ensure both that philosophical inquiry can access its subject matters and that sensitivity to data makes such inquiry likelier to be on track than otherwise. It is unclear that this tension can be satisfactorily resolved.

This brings us to the second point: while there may be ways to resolve the tension, they risk simply moving the bump in the rug by generating incoherence elsewhere. To appreciate this, suppose we were to cross the gap by endorsing a bridge principle that licenses a transition from the relevant psycho-linguistic claims to the correlative (not exclusively psycho-linguistic) subject matters; such a principle would facilitate access to the latter via the data. Suppose, for example, that a theory were to embrace the principle that things generally are as they appear, which would allow theorists to get at (say) the connection in Preferences via Preferences$_\psi$. This principle is transparently incompatible with the fundamental commitments of a number of philosophical theories, which decidedly reject this conservative principle in favor of a revisionist take on various philosophical matters. For instance, no version of nihilism about aesthetic properties could endorse it. To the contrary, these theories are committed to rejecting any such bridge principle, as they are committed to the claim that when it comes to matters of beauty, things tend not to be as they appear. Likewise for error theories about morality, or austere ontologies in metaphysics, or radical forms of instrumentalism in philosophy of science: they deny that things generally are as they appear. The challenge for proponents of the psycho-linguistic theory, then, is to identify a principle that bridges the gap between psycho-linguistic data and the subject matters of philosophy, but that isn't in tension with the commitments of their own philosophical views. It is unclear whether this challenge can be met (at least in a way that stays neutral between conservative and revisionary views).

The third point follows upon the second. The challenge just described arises from the psycho-linguistic theory's posit of a sizeable gap between the philosophical data and the subject matters of philosophy. The principal motivation for this posit is the aspiration to secure neutrality. But as we will endeavor to show in the next chapter, in the presentation and defense of our preferred view of data, neutrality can be achieved in another way. And so the psycho-linguistic theory, compelled to chase down an acceptable bridge principle, introduces a superfluous epicycle in philosophical theorizing.

So far we've assessed the psycho-linguistic theory in light of the data about data. We close with a pair of rather different observations about this theory, the first of which highlights its divergence from actual philosophical practice. In point of fact, nearly all prominent philosophical theories have recognized that the data are not exhausted by psychological or linguistic phenomena. Aesthetic theories, for instance, offer accounts not simply of Preferences$_\psi$ or Preferences$_\lambda$ but also of the connection specified in Preferences. And when criticizing objectivist views, subjectivists charge not that their rivals are unable to account for Preferences$_\psi$ or Preferences$_\lambda$, but rather that they cannot handle the connection specified in Preferences (or cannot do so as well as subjectivist theories do). A similar narrative applies in metaethics, where expressivists object not that moral realism fails to account for the appearance of, or felicitous speech about, the motivational efficacy of moral judgment, but rather that the view is unable to make sense of the motivational power of such judgment. These philosophers take the data to be not mere psychological or linguistic claims of the sort recognized by the psycho-linguistic theory, but rather to be features, or putative features, of the subject matter itself.[14]

---

[14] *Cp.* J. L. Austin's famous clarification of his preferred linguistic methodology, which in fact demands that theorists attend to the subject matter itself: "When we examine what we should say when...we are looking again not *merely* at words...but also at the realities we use the words to talk about: we are using a sharpened awareness

Our second observation is not descriptive but normative. The psycho-linguistic theory makes a mockery of much legitimate philosophical inquiry, which in the paradigm case explores the nature and status of the denizens of a non-psychological, non-linguistic reality. Philosophers are striving to answer such questions as whether God exists, or whether a priori knowledge is possible, or whether there are objective aesthetic and moral facts. We misconstrue the philosophical enterprise and are bound to fail to answer its central questions if we focus exclusively on claims about our psychological states or the language we happen to use to explore these questions.

We turn in the next chapter to presenting an alternative view of data, one that avoids the foregoing problems. The central idea is that what qualifies as a datum is intimately bound up with what claims inquirers have epistemic reason to believe. This theory gets the connection between data and the subject matters of philosophy right. It also handles all three of the basic features of data that trip up the competition. Moreover, we'll argue, it illuminates several other important features of data, such as their legitimate rejection under certain conditions and the process of data collection.

of words to sharpen our perception of, though not as the final arbiter of, the phenomena" (1956–7/1979, 182).

3

The Epistemic Theory of Data

Philosophical theorizing must start somewhere. According to the model of inquiry we've presented, the place to begin is with the data of the domain under investigation. We've argued that such data are not bound to be problematically theory-laden. Nor are they best construed in sociological, metaphysical, linguistic, or psychological terms. Rather, in our judgment, the correct theory of data is one that privileges their epistemic status.

We devote this chapter to a presentation and defense of our preferred version of this view. Our defense will cover all four basic features of data enumerated in the previous chapter: that data are starting points, inquiry-constraining, collected, and neutral. We'll take an especially close look at the process of data collection, identifying its constituent elements and showing how our version of the epistemic theory helps to illuminate its structure and dynamics. We'll also discuss how our view enables a neat formulation of the conditions under which data are legitimately rejected. If our

Philosophical Methodology: From Data to Theory. John Bengson, Terence Cuneo, and Russ Shafer-Landau, Oxford University Press. © John Bengson, Terence Cuneo, and Russ Shafer-Landau 2022.
DOI: 10.1093/oso/9780192862464.003.0004

contentions are correct, they fill a considerable lacuna in recent work in philosophical methodology, providing a fully general theory of data in philosophy.

1. The Epistemic Theory

According to the epistemic theory of data, φ is a datum regarding domain D if, only if, and because φ is about D, does not belong to any well-formed theory of D, and inquirers are in a good epistemic position with respect to φ.[1]

Different versions of the epistemic theory may explicate the relevant epistemic position in different ways. Of course, any acceptable explication must be capable of handling all four basic features of data. We take this to imply that a view that characterized 'good epistemic position' in terms of knowledge would be too strong, since knowledge is factive, and so not neutral. By contrast, a position that focused merely on epistemic permissibility would be too weak, since it would fail to ensure that data are inquiry-constraining; after all, such permission is realized by a great number of beliefs, including many with contents the neglect of which would not lead inquirers off-track. In between these lies the good epistemic position that we seek.

Although we are officially neutral regarding the correct explication of 'good epistemic position,' we will seek to reveal the attractions of the epistemic theory by employing the familiar category

---

[1] This formulation of the epistemic theory pertains to data as they figure in the *structure* of theoretical inquiry (recall Chapter 1, §1). Since the epistemic position of inquirers may change over time, characterizing the data at a particular moment in an actual instance of inquiry would require temporal indexing.

of an *epistemic reason for belief*. Four features make this a congenial candidate.

First, it is weaker than knowledge but more demanding than epistemic permissibility, and so avoids the aforementioned concerns about each. Second, it is a defeasible epistemic status—it is possible for reasons for belief to be outweighed or extinguished by sufficiently weighty countervailing considerations. When they are not, those reasons are undefeated but remain defeasible. Third, some reasons for belief are 'strong': they would not be easily defeated by competing considerations, and their possession implies the possession of (at least some) evidence. As will emerge below, we restrict our attention to strong epistemic reasons that are not defeated—we call them 'good' reasons. Fourth, the category of a reason for belief is not inherently individualistic but also countenances groups. On the view of data that we are developing, the relevant reasons are broadly social in the sense that they are reasons for inquirers *considered collectively*. The idea here is that when it comes to data, what matters are the good reasons possessed by a group, namely, the inquirers, rather than those reasons possessed by one or another individual.[2]

We call this approach to data the

Epistemic Reasons Theory: φ is a datum regarding domain D if, only if, and because φ is about D, φ does not belong to any

---

[2] We are willing to adopt the working assumption that a reason is possessed by inquirers, considered collectively, just in case a (possibly non-trivial) subset of the inquirers in the collective possess that reason, and it is available to most of the rest. This captures the idea that a group may possess a reason even when one or more of its members lack it (perhaps because they happen to endorse idiosyncratic views that function for them as defeaters of the reason possessed by the group, which possesses defeaters for those defeaters).

well-formed theory of D, and inquirers, considered collectively, have good reason to believe φ.[3]

Regarding the truism that data are considerations that must somehow be handled when theorizing about a domain, this theory implies that 'must' functions as an epistemic modal, serving to indicate what the balance of reasons requires.

While the epistemic reasons theory provides a substantive account of what data are, it is also non-committal in some important respects. It is compatible with different accounts of what it is for inquirers considered collectively to have reasons, as well as why inquirers have them. For example, it can be combined with views as divergent as virtue epistemology (which appeals to inquirers' competences), phenomenal dogmatism (which focuses on inquirers' seemings or presentations), and subjective Bayesianism (which invokes probability given inquirers' background beliefs). It is also compatible with different analyses, explanations, or measures of normative strength. For instance, an explanation of why some reasons are stronger than others might privilege modal features, such as resilience in the face of exposure to additional evidence, or emphasize etiological factors, such as being the product of dialogue with differently minded interlocutors.[4] Further, the epistemic reasons theory is compatible with

---

[3] Recall that a theory is 'well-formed' just in case it is a set of claims that satisfy a sound method's criteria at the second stage of inquiry. The conditions advanced by the epistemic reasons theory imply that each domain includes a very wide range of data. Those wishing to narrow the pool may add a fourth condition to the three already listed: φ is not handled by a theory of a domain other than D. Our notion of core data (in §2, below) points to a further way to zero in on the important elements of a data set.

[4] See, e.g., Jaggar (1989, §VII–XI) on the important role of "outlaw" and "subversive" insights in inquiry. Our understanding of strength is designed in part to protect against concerns about insular groups, as well as those dominated by bullies or poseurs.

different views about what is required to count as a defeater for a reason of the relevant strength. (We'll return to this last issue in the next section.)

Being both substantive and flexible are advantages of the epistemic reasons theory. But its primary virtue lies in the fact that it handles the four basic features of data. Begin with the observation that data are starting points. As it should, the epistemic reasons view implies that data are not the outputs of theorizing in that domain. If a set of claims satisfy the view's conditions, then those claims are about the domain, and inquirers, considered collectively, have good reason to believe them. Accordingly, these claims have a place among those that must be handled by a theory of the domain. In effect, they play the role of inputs to theorizing, rather than of outputs, per this first datum about data.

Unlike some of its rivals, the epistemic reasons theory also handles neutrality: the data function as common currency, while also being non-factive. They function as common currency because the good reasons to believe them are possessed by inquirers considered collectively. This approach fits with all three dimensions of common currency described in Chapter 2 (§1). Data merit attention even by theorists who might reject them because data are backed by strong reasons that, though defeasible, remain undefeated for the group. Data are often open-ended because the claims that inquirers considered collectively have good reason to believe are often vague or underspecified. And the data do not determine a unique theory because such claims are potentially compatible with a number of different views.

The epistemic reasons theory does not entail that the data hold or are true. So it smoothly handles the non-factivity of data as well. By contrast with the metaphysical theory of data, the epistemic reasons theory recognizes the potential for a gap between the data and the subject matters of philosophy. Yet unlike the psycholinguistic theory, the epistemic reasons theory denies that this gap is one between two disparate types of contents. Indeed, on the epistemic reasons view, when the data are correct, their contents match the contents of the domains that they are about.

The view also helps us to see how to bridge the gap. It does so via an epistemic treatment of data's status as inquiry-constraining: in a nutshell, a data set serves to *epistemically* link an inquiry to its subject matter (through a reason relation) in two ways. First, given that data are considerations that inquirers have good epistemic reasons to believe about a domain, the data pool operates as the principal means, or epistemic avenue, by which inquirers access their subject matter. Second, investigations of a domain that entirely ignore the data set are likely, or at least significantly likelier—epistemically likelier, given the good epistemic reasons inquirers possess—to be off-track with respect to that domain. If there weren't any epistemic reasons, or there were only trifling or defeated ones, then the inquiry wouldn't be properly anchored in its subject matter.

The epistemic reasons theory also makes sense of the idea that data are collected, by explaining central features of such collection. Procedures for data collection are often rightly subject to scrutiny. When presented with a body of data, we often want to know what procedures inquirers have used to gather data, and whether those procedures have the requisite credentials. According to the epistemic reasons view, this desire flows from the demand to use only the procedures that provide good *reason to believe* that the domain is one way rather than another. In the paradigm case, when collecting data, inquirers take their procedures to provide strong, undefeated reason to believe certain claims about the domain. For example, if inquirers think they should collect data from the outputs of our best scientific theories, that is because they hold that those outputs are ones they have good reason to believe. Or if inquirers think they should collect data via intuitions about thought experiments, that is because they hold that such intuitions provide good reasons for belief. The epistemic reasons theory straightforwardly accounts for this dimension of data collection.

It also explains at least two types of controversy over candidate procedures for data collection. One type of controversy targets specific procedures, with one set of theorists endorsing a certain procedure while their opponents favor some alternative. Take, for example, debates among friends of intuition about whether intuitions regarding general principles or concrete cases should be endorsed when they conflict.[5] The epistemic reasons theory reveals what is at issue in such disagreements, namely, the comparative strength of the epistemic reasons that are generated by each class of intuitions.

A second type of controversy occurs at a somewhat more abstract level. Some philosophers might insist, for instance, that to be a legitimate procedure, we must have independent warrant for regarding its verdicts as reliable.[6] Others would reject this. These healthy disagreements make sense if, per the epistemic reasons theory, a procedure for data collection functions to provide good reason for belief. The controversy arises because the

---

[5] Bealer (1992, 104) privileges intuitions about concrete cases, while Huemer (2008) defends the opposite view in the case of ethics. *Cp.* Weatherson (2003).

[6] *Cp.* Cummins (1998), Wright (2004), and Weinberg (2007).

question of what it takes to satisfy this condition is both open and often extremely difficult to answer.

The same is true of the question of *which* claims satisfy the condition. Thus the epistemic reasons theory sheds light on disputes over which claims are to be counted as data. Such disputes are real, and can be important. If our view is correct, these disagreements are predictable. They can be traced to an unfortunate but familiar opacity that sometimes attends our efforts to determine where the balance of reasons lies.

Some would try to make things easier by restricting philosophical data to a specific set of considerations, such as those that are widely accepted, reflected in ordinary usage of language, or scientifically verifiable. The epistemic reasons theory delivers the (correct) verdict that these views are guilty of chauvinism. For they are illicitly denying that any other considerations could supply good epistemic reason to take the relevant domain to be one way rather than another. We can avoid such chauvinism by viewing data as epistemically well-supported considerations, whatever those happen to be.

To reject chauvinism is not to suggest that anything goes. The epistemic reasons theory sees philosophical data as delimited by the good reasons possessed by inquirers, considered collectively. It also clarifies what it takes to precisify a datum. A satisfactory precisification of a given datum φ must be a claim backed by good epistemic reason, while not omitting anything in φ that is thus supported. In this way, the epistemic reasons theory illuminates the propriety of precisifications.

The epistemic reasons theory also bears on the propriety of methods. Given that inquirers have good reason to take data to be genuine features of the domain under investigation, no method of theorizing could qualify as sound were it insensitive to the evidence that data provide. Likewise for any method that allows theorists to reject the data without justifying their rejection. In effect, our view explains why the data have a profound influence on what qualifies as a sound method.

Finally, consider the character of theoretical inquiry itself. Theoretical inquiry is not an epistemically neutral process. Rather, when inquirers partake in it properly, they enjoy a certain kind of good epistemic standing: they have good epistemic reason to believe a range of claims about the domain. Notice the Moore-paradoxical character of the following statement, uttered by inquirers in a given domain: "These are data, considerations that must be handled, but we lack reason to take them to identify genuine features of the domain." The epistemic reasons theory neatly explains not only why such a statement is as puzzling as it is, but also why properly engaging in inquiry is incompatible with the absence of any reasons for belief.

2. The Process of Data Collection

Philosophical data do not magically appear in ready-made lists; they must be collected. The epistemic reasons theory helps to illuminate the inner workings of such collection, and accords with best practices for data collection undertaken by individuals or groups.

As we see it, the process of data collection has three basic components. The first comprises its inputs, which include whatever is potentially relevant to answering the central questions about a domain—for example, actual or hypothetical cases, a linguistic corpus, survey results, and the like. The second comprises the outputs of data collection, which are the data themselves. The third component is a *procedure* that takes collectors from the former to the latter.

Procedures for data collection are ways of sifting and winnowing so as to determine what does and does not qualify as a datum in the domain at hand. Examples of procedures that have played prominent roles in the collection of philosophical data include:

- Perception of items in one's environment
- Introspection of one's own phenomenal states
- Intuitions about thought experiments
- Linguistic judgments about the grammaticality, meaning, or truth-conditions of sample sentences
- Common sense verdicts about what exists, is known, etc.
- Imagining future possibilities
- Inductive inference from observed samples
- Sociological observations about ordinary conversations
- Attending to the testimony of members of marginalized or oppressed groups
- Statistical analyses of survey results
- Selecting, implementing, and analyzing a model of a scenario
- Searching for the origins of a practice through archival study
- Interpreting a text using particular hermeneutics

It goes without saying that these procedures represent only a tiny sample of those that are and could be used to collect philosophical data. Note, also, that they can overlap, intersect, and be conjoined. We make no claim to either exhaustiveness or exclusivity.

We think of a procedure as the conjunction of two elements. First, there is a *source*: a way of forming thoughts about the various inputs, by means of which collectors arrive at a belief to the effect that the domain is a certain way. Candidate sources, featured in our examples just above, include perception, introspection, intuition, linguistic judgment, common sense, imagination, and inference. Second, there is a *technique*, which centrally involves the application of a source to an input. Candidate techniques, implicit in the preceding examples, include directing attention to one's environment, focusing on one's phenomenal states, reflecting on thought experiments, mulling over sample sentences, and so forth.

Sources and techniques can be mixed and matched. To illustrate, reflection on thought experiments will sometimes utilize intuition; at other times, such reflection will be allied with linguistic judgments concerning "what we would say." As this shows, one and the same technique can be combined with different sources, thereby yielding distinct procedures. Similarly, we can conjoin one source with different techniques, thereby creating additional procedures—as when we employ intuitions to issue a verdict about a thought experiment or instead to assess the plausibility of a general principle.

Sources must be in good epistemic standing. For if they weren't, then reliance on them would fail to provide collectors with good reasons for belief, thus violating the strictures imposed by the epistemic reasons theory of data. But reliance on sources in good epistemic standing is not enough to constitute a successful procedure for data collection. Promising sources must be utilized in tandem with promising techniques. While techniques may not be directly assessable for their epistemic merits, they can be evaluated with respect to whether they sustain, or instead compromise, the good epistemic standing of the source with which they are combined. Techniques possess a variety of properties that must factor into such evaluation: they can be simple or complex, efficient or inefficient, flexible or rigid, indiscriminate or selective. Some techniques will score well, while others are bound to thwart the effort to gather genuine data. The candidate techniques noted above arguably fall on the former side of this divide; reading stars, tea leaves, or Tarot cards, even when allied with a good source such as perception (of the stars, leaves, or cards), falls on the latter.[7]

In our view, there are a wide variety of legitimate combinations of sources and techniques (procedures), including many of those listed above. At the same time, we recognize that, even bracketing the epistemic duds, different procedures will sometimes provide reasons for belief of differing strengths. We thus affirm a discriminate pluralism, which recognizes multiple procedures possessing varying degrees of authority. One reason to favor this moderate position in philosophy is that philosophical data will probably contain both empirical and a priori elements (hence, pluralism), and it is plausible that some procedures will be epistemically better than others (hence, discriminateness).

To illustrate our pluralism, consider the variety of data operative in recent debates over the nature of know-how, a paradigm example of which is knowledge how to dance. Gilbert Ryle famously claimed in *The Concept of Mind* that know-how is not a species of propositional knowledge, but rather is a type of ability.[8] Ryle is often regarded as having developed his view in light of data collected using reflection on ordinary language. Data amassed using

---

[7] Contrast an approach that requires sources and techniques to satisfy the thesis we labeled 'Strict Neutrality' (Chapter 2, §2) or "produce consensus over time" (*cp.* Armstrong 1973 and Lycan 2019, 2 and 86).

[8] Ryle (1949, Ch. 2).

Illustration 3 Knowledge how to dance.

procedures that incorporate techniques from contemporary linguistics, pertaining to the syntax and semantics of embedded questions, have been invoked in favor of the opposite perspective.[9] So, too, have data gathered via statistical analyses of surveys (a procedure popular within experimental philosophy).[10] Other data have been collected through reflection on thought experiments.[11] Still more have been discerned through phenomenological reflection[12] and the interpretation of various findings in cognitive science.[13] The result of these diverse efforts at collection is an incredibly rich data set, supported by various types of evidence, which competing theories of know-how rightly aspire to handle.

---

[9] Stanley and Williamson (2001). In reply, Rumfitt (2003) and others have collected cross-linguistic and so-called corpus data that, they maintain, point in the opposite direction of Stanley and Williamson's position.

[10] Bengson, Moffett, and Wright (2009).

[11] See, e.g., Bengson and Moffett (2011) and Cath (2011).

[12] Dreyfus (2014).

[13] Glick (2011, §1) provides a helpful summary.

We do not take a stand here on the comparative merits of specific procedures, or on whether their promise will vary depending on the philosophical domain under investigation. Nor will we scrutinize traditional procedures or the ways in which they might encode systemic prejudices or biases. Our aim is not to settle such matters, but rather to reveal the structure of stage one of inquiry, and to show that the epistemic reasons theory offers a helpful perspective on the inner workings of data collection. Accordingly, we leave room for more detailed discussions about the viability of specific procedures, which our account of stage one helps to discipline. For the distinction between sources and techniques reminds us to keep separate the merits (or demerits) of one element of a procedure from those of another, while the epistemic reasons theory explains why the relevant merits (or demerits) are epistemic ones.

If pursued indefinitely, data collection would likely yield indefinitely many data. Since working with such a pool would be impractical, there are compelling reasons for theorists to seek out a relatively small data set. This can be done by bracketing those data whose handling would shed little light on the central questions in a given subfield. Foremost among these are miscellaneous data subsumed by other data. A domain's 'core' data are not like this. For an example of a datum that does not belong to the core, consider the claim that sheltering the homeless tends to contribute to a good life. Although this is a prime candidate for a datum about the good life, singling it out would be relatively unhelpful. After all, it's also a datum that active engagement in rewarding activities tends to contribute to a good life, and whatever considerations handle the latter datum would presumably do the same for the former, though not vice versa.

Core Data in Contemporary Metaethics

Since each domain contains a very large number of core data, the threat of cherry-picking is genuine. It is good practice for inquirers to occasionally take inventory, ensuring that their theorizing at stage two is properly moored.

To illustrate what this might look like, consider contemporary metaethics, whose central questions include "Is morality objective?", "How is moral knowledge possible?", and "Why be moral?" Inquirers in this domain recognize a wide range of data spanning the normative, metaphysical, epistemic, psychological, and semantic dimensions of morality. Some of these qualify as core data. For example:

- *Correlation*: How things are morally is correlated with how things are non-morally.
- *Independence*: Some moral demands hold regardless of what the subjects of those demands think, feel, desire, hope, fancy, etc.
- *Inescapability*: Some moral demands and reasons are inescapable with respect to normative assessment.
- *Priority*: Moral demands and reasons sometimes have priority over non-moral ones.
- *Non-Arbitrariness*: Morality is not arbitrary.
- *Non-Contingency*: It's not deeply contingent that some things have the moral features they do.
- *Practicality*: Moral judgments have marks of practical attitudes: they direct, guide, and motivate action.
- *Declarative*: Moral judgments have marks of declarative attitudes: they are predicative, truth-evaluable, and apt for inference.
- *Grasp*: Moral agents sometimes grasp, or are duly placed to grasp, moral reality.

- *Fallibility*: Moral agents can be mistaken about moral reality.
- *Evident*: Some moral claims are evident, sometimes highly so.
- *Agreement*: There is widespread moral agreement about a range of moral issues.
- *Disagreement*: There is widespread moral disagreement, some of which may be unresolvable.

These data do not exhaustively characterize the metaethical dimensions of moral thought, discourse, and reality. Nor are they fully precise, since they—like nearly all data—are fairly vague and open to clarification or explication. For example, they do not wear on their face their scope or modal force. Given that data are *inputs* to theorizing, this is as it should be. As neutral starting points, they can be approached in different ways, and further reflection may reveal precisifications that are considerably more determinate. Quarrels about such matters are fully compatible with recognizing all of the above as core data in metaethics.

Even without an algorithm for identifying core data, we can enlist various heuristics to assist us in this project. For example, inquirers can check whether a datum in domain D is saturated with minutiae, or is also a datum in many other domains besides D, or is entailed by another datum in D; any of these would serve to defeasibly indicate that the datum is not among the core in D. Alternatively, inquirers can ask whether a datum has played a central role in a philosophical debate concerning a given domain; its doing so would signal that it is a core datum therein. This conservative heuristic is counterbalanced by a more progressive one: should it turn out that a datum has been overlooked by those inquiring about a given domain, but its handling promises to transform debate regarding that domain, then this suggests that it is a core datum therein. Some procedures are likely to be better than others at implementing such heuristics, and so pinpointing a domain's core data. In this way, a lot can ride on which procedures are selected for use in the process of data collection.

3. Data Imposters, Data Disablers

We've emphasized that data play a central role in the construction and evaluation of theories. Because they are neutral starting points that constrain inquiry, no theory can simply dismiss or ignore them. But as we've emphasized, data are not inviolable, and we are fallible collectors of data. So any sound method of philosophical theorizing must allow for two possibilities.

The first is that of exposing a putative datum as a mere pretender, a datum in name only. Doing this is a matter of discovering that there is not, despite appearances, a good reason to believe a given claim, at least among inquirers considered collectively. Such claims are not data but imposters. They can be unmasked in myriad ways.

Inquirers might be reminded of a flaw in the procedure they used when collecting data, which inadvertently issued a counterfeit. Or it may be that they have for some time possessed evidence sufficient to rebut a claim they wrongly treated as a datum. Perhaps, for example, they had taken it for granted that a given claim is metaphysically possible, but now notice that this would contradict a proposition they have long known to be true. They might also see that there is a distinct thesis in the vicinity that they have good reason to believe (e.g., that the original claim is conceptually possible); it, but not the original, is a datum. This suggests a third scenario, which involves inquirers realizing that they had mistakenly taken their procedure to ratify one claim, when it had actually delivered good reason to believe another. This could happen if (say) what looked to be a weighty reason to believe a non-conditional claim turns out only to support a conditional whose consequent is that claim.

Though scrutinizing candidate data with an eye to unmasking imposters is a healthy methodological undertaking, it is possible to be overzealous in this regard. Among other things, skeptical fervor can lead to illicit dismissal of genuine data. We can elaborate on our last example to illustrate the point. Consider the putative datum in normative ethics and moral psychology that there are moral exemplars who possess virtues such as beneficence, humility, and wisdom. Those who doubt the existence of such exemplars might insist that the real datum is that *if* anyone is a moral exemplar, then she possesses virtues—this conditional, rather than its non-conditional consequent, is what we have good epistemic reason to believe.[14] Such a strategy may be correct in some cases. However, it bears noting that there is sometimes good reason to believe a given claim *and* its correlative conditional. For what it's worth, that is what we're inclined to say in the case of moral exemplars. Whether or not that opinion is correct, the point is simply that when hunting for a claim to replace a putative datum,

[14] Skeptics include Harman (1999) and Doris (2002); their early critics include Sreenivasan (2002) and Kamtekar (2004). Our comments here speak to concerns that error theorists, nihilists, and other revisionists about various domains might harbor regarding how a data set can be inquiry-constraining, despite containing a multitude of non-conditional elements that those theorists deem to be false.

care is required to determine exactly which claims inquirers have good epistemic reason to believe.

Turn now to the second possibility, which concerns not imposters but disablers, by which inquirers legitimately reject a genuine datum. Our aim here is to countenance scenarios in which a given claim was a datum, or perhaps still qualifies as such, though some theorists *justifiably* endorse views that contradict it. Data disablers make these scenarios possible. Methodologically, it is important to register the conditions under which something counts as a data disabler, justifying the rejection of a given datum. The epistemic reasons theory specifies these conditions.

The challenge is to identify conditions that avoid two extremes. On the one hand, it shouldn't be too easy to qualify as a data disabler. For example, data are not rightly rejected just because they conflict with one's pet theory, or because addressing them would be complicated. On the other hand, it shouldn't be too difficult for theorists to identify data disablers, since inquiry shouldn't be biased towards conservative conclusions.

With the epistemic reasons theory of data in hand, we can identify the following conditions under which a set of considerations C qualifies as a data disabler:

C disables a datum for a set of inquirers if and only if C provides them with good reason either to believe that the datum is mistaken or to suspend judgment with respect to it, and this reason is at least as strong as the reason(s) for believing that datum.

When the good reason in question is possessed by inquirers considered collectively, the claim loses its status as a datum. But there are also cases in which the good reason is possessed not by inquirers considered collectively, but only by (say) the cognoscenti.

In such cases, the datum retains its status as such—it is still the case that inquirers considered collectively have good reason to believe it—even while some theorists legitimately reject it.

The biconditional above is both substantive and informative. But it stands in need of further specification in at least one respect: it does not indicate the conditions under which one reason is as strong as another. Here, as elsewhere, we seek to avoid committing ourselves to a specific (and probably contentious) position, leaving the resolution of this detail for further investigation. Still, even without having specified necessary and sufficient conditions for such comparative strength, we can make five comments about what sorts of considerations are and aren't good candidates for being data disablers.

First, suppose that a set of inquirers encounter new evidence that a procedure is epistemically defective in a particular respect, evidence that is at least as strong as the reason they previously had to believe its outputs. Or suppose that a set of inquirers uncover fresh evidence that decisively supports a position that contradicts one of their data. In both cases, such evidence functions as a data disabler.

Second, suppose a consideration consists exclusively in, or is fully epistemically based on, data from another domain, or elements of our best picture of the world. (For present purposes, think of this as including the deliverances of our best sciences and certain elements of common sense. In Chapter 5, we'll offer a somewhat more exact characterization.) If such a consideration casts serious doubt upon a datum, then it is a good candidate for being a data disabler. For such a consideration would tend to be in very good epistemic shape, thereby providing sufficiently strong reason to hold that a given datum is mistaken. After all, what most ordinary agents have reason to believe may be decidedly out of step with elements of our best picture of the world—in particular, with the deliverances of our best science. If these elements conflict with a datum that is backed by reasons no stronger than the reasons supporting those elements, then we have the makings of a data disabler.

Third, suppose a claim conflicts with some data but is such that its endorsement by a theory would thereby increase the theory's virtuosity—for instance, its simplicity or elegance. That by itself would not suffice to make the claim a data disabler. The claim that there are no group intentions might promise theoretical riches for any view that embraced it, insofar as such a view would be more parsimonious than one that affirmed the existence of such intentions. But this does not thereby imbue that claim with the type of epistemic significance required to disable the datum that (say) the Trump administration intended to promote skepticism regarding the legitimacy of the 2020 presidential election.

Fourth, and more tentatively, if a consideration is, or is fully epistemically based on, a theory that purports to answer one of the central questions of philosophy, then at least with respect to a wide range of data, it is *not* likely to be a good candidate for being a data disabler. Take, for example, the no-self theory of personal identity, endorsed by some Humeans and Buddhists. Whatever the view's merits, it is unlikely by itself to possess the stout epistemic profile required to disable the datum that some behaviors are accompanied by experiences of personal agency. The point applies equally to theories such as physicalism, dualism, idealism, platonism, nominalism, naturalism, theism, atheism, hylomorphism, hedonism, skepticism, or the like. To be sure, views of these sorts can play important roles in the construction and evaluation of theories. And when sufficiently specified, some of these views might be incompatible with data from one or another domain. But even when sufficiently specified, it is doubtful that any of these theories enjoys the lofty epistemic standing needed to function as a disabler regarding a great many data across philosophy.

The point we're making extends beyond philosophy to grand theories in other intellectual fields. The attachment theory of relationships (psychology), structural functionalism about societies (sociology), revealed preference theory (economics), the selfish gene theory (biology), and the Copenhagen interpretation of the quantum realm (physics)—each may play a role in inquiry, even while lacking the august epistemic credentials required to serve as a data disabler.

This is not to say that it is *impossible* for such theories to earn that status, providing good epistemic reasons that are at least as strong as the reasons favoring various data. Our main thought is simply that we currently have reason to think that theories purporting to fully answer one of the central questions of philosophy or another intellectual field are unlikely candidates. There might be other views in better epistemic standing than those just mentioned. Perhaps such views as S4 modal logic, causal realism, or the theory of natural selection have good enough epistemic credentials to play the role of a data disabler. We take no stand on these possibilities. This leads to our fifth and final point: the question of whether positions such as these are good candidates for being data disablers will often need to be assessed on a case-by-case basis. By specifying conditions under which data are disabled, the epistemic reasons theory paves the way.

Data are the lifeblood of philosophical inquiry, neutral starting points that serve as inputs to theorizing. Given their vital role, we must use care when gathering data, taking pains to exclude imposters. And even after we've ratified a datum as the genuine article, we must be alert to potential disablers. Be that as it may, no datum can be dismissed lightly, and it is perilous to reject all of the data. For data are inquiry-constraining, anchoring inquiry to its subject matter. The epistemic reasons theory makes good sense of these (and other) observations. As we turn to questions about theorizing, we'll want to remember the numerous benefits of the view. That said, many of the positive claims that we will make about stage two of inquiry are independent of this particular conception of data. Still, they form a natural package—a picture of methodology that improves on extant trends and approaches, several of which we'll explore in the next chapter.

4

The Question of Method

We turn now to method, the engine of inquiry. While the first stage of inquiry consists in the collection of data, the second centers on the construction and evaluation of theories. We call the latter stage—the topic of this chapter and the next—'theorizing.'

Chapter 1 surveyed the three constituent elements of theorizing: its inputs, its outputs, and the method that takes theorists from the former to the latter. The inputs to method are data. The outputs of method are theories. Methods themselves comprise a set of criteria that serve a dual role: they provide instructions for the construction of a theory, given the data, while also serving as standards by reference to which the merits of theories are evaluated. For example, a method might involve criteria instructing theorists to craft views that are internally coherent, implied by the data, and simpler than rivals. These criteria would, in turn, function as means by which theories are assessed.

The principal measure of a method's success is its ability to position inquirers to achieve inquiry's ultimate proper goal. What

Philosophical Methodology: From Data to Theory. John Bengson, Terence Cuneo, and Russ Shafer-Landau, Oxford University Press. © John Bengson, Terence Cuneo, and Russ Shafer-Landau 2022.
DOI: 10.1093/oso/9780192862464.003.0005

we label the 'question of method' asks us to identify a method that will take theorists from data to a theory that realizes such a goal.

The Question of Method

The question of method arises in part because, as we've noted, the data do not by themselves select a single theory, let alone one primed to resolve inquiry. To answer the question of method is, among other things, to address such underdetermination of theory by data. For it is to specify what enables inquirers to home in on a theory that is not simply compatible with the data, but realizes inquiry's ultimate proper goal—theoretical understanding.[1]

Some thinkers believe that there is no answer at all to the question of method: such skeptics deny the existence of a method that will, in suitable conditions, lead philosophers who follow it from the data to understanding. Others think that there is no *principled* answer to this question: such anarchists contend that the best strategy for attaining inquiry's ultimate proper goal is simply for whoever to do whatever, however, whenever, wherever, whyever.[2] We disagree, and in this chapter and the next will examine views with greater methodological ambitions.

---

[1] Recall that our use of the phrase 'inquiry's ultimate proper goal' should not be read as implying uniqueness, but rather as shorthand for the claim that theoretical understanding, but no weaker epistemic or alethic achievement, suffices to resolve theoretical inquiry (see Chapter 1, §3).

[2] Feyerabend's *Against Method* (1975) is the *locus classicus* of anarchism. Some anti-theorists gravitate toward the skeptical position described in the text (see, e.g., Horwich 2012, Ch. 2).

Our discussion begins by identifying a pair of data about philo-sophical method that will be familiar from paradigm examples of philosophical theorizing. Through a careful look at some specific examples of actual philosophical practice, we'll then identify a third datum centering on virtues of a sound method—one such that successfully following its instructions positions inquirers to achieve theoretical understanding. These exercises will clarify the question of method and uncover the resources needed to explain why extant philosophical methods do not adequately answer it. This will set the stage for our presentation, in the next chapter, of an alternative method that draws from familiar approaches, while seeking to provide that answer.

1. Two Data about Method

Our claims about method have to this point been highly sche-matic. We've said that its inputs are data: considerations that must be handled by its output, a theory. We've also emphasized that a method will not simply acknowledge that there are such data, but will provide specific criteria for theorists regarding exactly what it is to handle them. In effect, these criteria issue norms of inquiry that govern its second stage. By 'method,' then, we do not simply mean a tool for doing philosophy—like the method of cases, or the phenomenological method, or genealogical critique.[3] Rather, we mean something that specifies what to do when transitioning from the data to a theory.

---

[3] There are of course many other types of tools. See, e.g., Maitra's (2016, §6) list of eight "tools that have been central to feminist work in philosophy," and Sider's (2020, Ch. 1) inventory of modal and post-modal "tools in metaphysics."

We're now ready to put flesh on these bones, by collecting some data that will anchor our search for an answer to the question of method. A good place to start is by reflecting on standard philosophical practice, observing that when they endeavor to assemble a theory, philosophers[4]

- advance arguments
- raise objections
- offer replies to these objections
- provide clarification
- develop explanations
- display sensitivity to the deliverances of logic, mathematics, science, and (to a greater or lesser degree) common sense.

Whatever philosophical method is, it is something that is *friendly* to these activities. By this we mean that, in the paradigm case, implementing philosophical method involves engaging in such activities. This, we believe, is a datum about philosophical method (hereafter, simply 'method').

The claim that method is friendly to the above activities is an inquiry-constraining starting point that is common currency among those inquiring about method. Like other data, this 'friendliness datum' is also non-factive: it may be mistaken. Nevertheless, it is fair to say that a set of criteria for theory construction and evaluation in philosophy that is *unfriendly* to one or more of the activities listed above is best classified as either incorrect, incom-

---

[4] By 'philosophers,' we mean those engaged in philosophical inquiry, understood as a species of theoretical inquiry. And below, when we speak of 'doing philosophy,' we mean engaging in such inquiry. As emphasized in the Introduction, we recognize that philosophy can be done (and done well) in many other ways.

plete, or revisionary. For such criteria will comprise a method that fails to countenance these activities. And, as should be clear, we have good epistemic reason to think that engaging in these activities is part of doing philosophy.

Illustration 4 Socrates, Aspasia, Pericles, and Alcibiades in debate.

The friendliness datum may seem obvious, but it is worth making explicit. For it ensures that an investigation of method properly connects to actual philosophical practice. And as it happens, some views have trouble handling it. Consider, for instance, simple methods that call on theorists to construct a theory that merely fits the data—for instance, one that is *implied by* the data, or instead *probabilifies* them. Although we take such methods to contain a kernel of truth, the criteria they advance do not fairly represent philosophical practice. While crafting a theory that is merely implied by or probabilifies the data is compatible with engagement in most of the above activities, such approaches could just as easily be combined with a "trial and error" strategy that eschews the various elements of philosophical debate enumerated above. In effect, the simple methods fail to make sense of the friendliness datum. They are too simple.

Unsurprisingly, then, these views cannot satisfactorily answer the question of method. For if it is a datum that philosophical method as such is friendly to the activities described above, then it is also a datum that a sound method will also be friendly to such activities. Put differently, the correct answer to the question of method is one that will render it likely that when transitioning from data to a theory that is poised to deliver theoretical under-standing, philosophers will engage in the above activities. But the simple methods do not do this.

Such methods also fail in another, arguably deeper respect. Call a method 'determinative' only if its criteria select among multiple theories all of which fit the data, ruling a great many of them out of contention. Determinativeness addresses the underdetermination of theory by data. To the extent that implication, probabilification, and other such relations are little more than ways that multiple theories could all fit the data, the simple methods lack criteria that take the next step. They are not determinative, and so fail to be sound.

Two important features of a sound method, then, are that it will be friendly to the above activities and that it will be determinative in the sense just specified. But the friendliness datum and what we'll call the 'determinativeness datum' aren't the only data that must be handled by an answer to the question of method. Next, we'll try to collect a third, centering on a constellation of features exhibited by a sound method. Doing so will position us to take a careful look at four methods that have dominated contemporary philosophy.

2. A Third Datum

Our strategy for identifying this constellation of features is indirect. We'll isolate a series of methodological tendencies through critical examination of classic and contemporary philosophical work. Many of these tendencies are laudable, and combine to form a multifaceted datum regarding a sound method. However, they're closely associated with and can easily devolve into tendencies that are problematic. Our task is to disentangle them.

2.1 Methodological trends in recent philosophy

We begin with an observation concerning philosophers' tendency to sniff out potentially troubling aspects of rival theories. While it is laudable to engage with other positions in ways that seek to identify their shortcomings, such engagement becomes problematic when it leads philosophers to conclude that the target of their objection is inferior to their own position, absent the comparative work required to legitimate such a conclusion.

Hartry Field's scrutiny of mathematical realism's commitment to the existence of mind-independent numbers is commendable, as it rightly highlights the fact that mathematical realists face the serious challenge of explaining how we could grasp numbers so understood.[5] Arguably, however, theorists inspired by Field's discussion would take a wrong step were they to conclude that realism is inferior to anti-realism just on the basis of the allegation that realism fails to explain the reliability of our mathematical beliefs. To establish such a conclusion would require pursuing the question of (say) whether anti-realism suffers from problems as severe as those that are thought to beset realism.

A second laudable tendency in philosophical inquiry is to identify central features of a domain, and to insist that a theory of that domain should have something to say about what they are and why they obtain. However, attention to such phenomena sometimes encourages theorists to discuss features of the domain piecemeal, restricting their analyses to certain types of considerations, which are treated as if they decisively establish one or another theory regarding that domain.[6]

Sharon Street's discussion of normative realism displays admirable sensitivity to evolutionary considerations pertaining to the origins of normative thought. But it also risks falling prey to an illicit temptation when it claims that these considerations themselves "settle" the debate against realism.[7] After all, properly assessing the merits of philosophical theories for their capacity to facilitate understanding of their targets requires taking into account not just a single consideration, or type of consideration, but a wide range of data.

[5] Field (1989, 25–30 and 230–9; 2005). Field describes his critique of realism as a "reconstruction" of an argument presented by Benacerraf (1973).

[6] *Cp.* Wittgenstein (1953, §593): "A main cause of philosophical disease—an unbalanced diet: one nourishes one's thinking with only one kind of example."

[7] Street (2008, 207 and 2015, 692).

A third, rather different laudable tendency consists in recognizing the importance of defending one's theory in the face of various challenges. In *The View from Nowhere*, Thomas Nagel candidly acknowledges and addresses several influential arguments against the normative realist position that he endorses. Appreciation of the necessity of defense should, however, be distinguished from the unfounded supposition that there is nothing more, beyond replying to such objections in a coherent way, that must be done in order for one's view to merit allegiance over rival theories.

Nagel flirts with a supposition of this kind when he tells us that he need not—indeed, *cannot*—say anything to "render [realism] more plausible than the alternatives," beyond pointing to an initial, unearned "presumption that there are real values and reasons" and noting resources to "refute impossibility arguments."[8] However, while it may be that some beliefs (e.g., basic perceptual or introspective ones) enjoy a standing presumption in their favor, there is little to be said on behalf of the idea that a *whole philosophical theory*, such as normative realism, requires no positive support whatsoever. Insofar as a theory purports to provide genuine understanding of—and not merely (say) a permissible perspective on—some domain or subject matter, its claim to do this best must be earned, not presumed.[9]

Philosophers have been sensitive to the potential theoretical interest of distinctions among and analyses of central philosophical terms, concepts, or properties—witness twentieth-century epistemology's obsession with the analysis of knowledge in the wake of Edmund Gettier's famous counterexamples to the thesis

---

[8] Nagel (1986, 143–4).

[9] *Cp.* Korsgaard (1996, 41–2), who raises a similar worry.

that justified true belief is sufficient for propositional knowledge. Such sensitivity is a fourth laudable tendency. But distinction and analysis are not all there is to constructing and evaluating philosophical theories. To borrow terminology from H. H. Price, besides "analytic clarity" there is also "synoptic clarity," which consists in "bring[ing] out certain systematic relationships" among various elements of a given domain.[10] The paradigm of such a relationship is explanation, but others, like coherence, are also bound to be important. Thus when epistemologists assert or assume that "What is knowledge?" is the principal question of their field, they not only run the risk of illegitimately downplaying a variety of other epistemic phenomena, such as the character of intellectual virtues, epistemic reasons, and understanding. They also risk neglecting the importance of providing synoptic clarity regarding knowledge itself, including the systematic relationships it bears to such things as inquiry, wisdom, expertise, and the good life.

Other tendencies in contemporary philosophy center on the variety and strength of the criteria by which theories are evaluated and their comparative merits determined. Two such tendencies are clearly commendable. The first, concerning strength, consists in acknowledging that certain criteria take priority over others. Ted Sider, for instance, endorses the comparative significance of parsimony when he argues for the nihilist thesis that there are no composite objects, on the grounds that "nihilism allows us to eliminate 'part' from the ideology of our fundamental theories."[11] The second, concerning variety, consists in recognizing a plurality of criteria, such as explanatory power, coherence, and simplicity. Particularly noteworthy is David Lewis' sensitivity to a wide range

[10] Price (1945, 29).   [11] Sider (2013, 240).

Illustration 5 An alleged composite object disassembled.

of disparate criteria (such as fit with common sense, unity, and economy), on display in his attempt to defend the plausibility of his concrete modal realism.[12]

Such sensitivity sometimes tempts theorists to treat several criteria as having the same importance as every other. But not all criteria are created equal, as the laudable tendency to prioritize some criteria implies. However, this tendency can also lead theorists astray. Consider the practice of elevating just one criterion above all others, and rendering a judgment about the comparative merits of different theories by reference to the privileged criterion alone. The scientistic allegation that a theory's empirical bona fides solely

---

[12] Lewis (1986, 3–5 and 134–5). See also Lewis (1983, x–xi); cp. Lewis and Lewis (1970, 211–12) and Armstrong (1989, §1.4).

determine its viability is one example. Treating parsimony as not merely important but as invariably overriding is another, since parsimony is not always on the side of the good.[13] If theory T is more parsimonious than theory T*, but T's parsimoniousness makes it *less* likely to promote understanding than T*, then T's parsimoniousness does not win it *any* credit, much less any that could compensate for shortcomings elsewhere. More generally, a view cannot be entirely discounted by its doing less well than rival theories with regard to just one criterion. Nor is the mere fact that a theory fares better according to one dimension of evaluation sufficient to make it best (i.e., most likely to further understanding of its subject matter). A theory's status as such is rightly determined only after the application of multiple criteria.

2.2 The rudiments of a sound method

Our analysis has uncovered a set of laudable methodological tendencies. Each corresponds to a methodological virtue, insofar as it is a feature that enables a method to position inquirers to achieve inquiry's ultimate proper goal. If our discussion hits the mark, a sound philosophical method must be:

- *Comprehensive*: it calls for theories to address not just a single consideration, or type of consideration, but a wide range of data.

[13] Philosophers such as Quine (1948), Lewis (1986, 4), Churchland (1984, 18), Gibbard (2003, xii), Melnyk (2003, 244ff.), and Olson (2014, 147–8) have shown sympathy for the view that parsimony is invariably overriding. Distinguish the assumption that one criterion reigns supreme from the reasonable position that there are certain necessary conditions on the plausibility of theories (e.g., handling the data in the domain). One can endorse the latter position without at the same time assigning a single criterion overriding importance with respect to all others.

- *Support-requiring*: it delivers a theory that is epistemically well-supported by positive considerations that speak on its behalf.
- *Synoptic*: it ensures that whichever theory it favors is best positioned to expose systematic relationships, including explanatory ones, in the subject matter in question.
- *Multidimensional*: it incorporates multiple criteria.
- *Hierarchical*: it recognizes that criteria play diverse roles, and that some roles possess more significance than others.
- *Comparative*: it enables assessment of both individual and relative merits of rival theories.

That a sound method exemplifies these many virtues is a neutral, inquiry-constraining starting point when thinking about method. It merits attention from proponents of disparate candidates for a sound method, is well-suited to anchor their theorizing about this matter, and is not itself an output of such theorizing. So, despite its complexity and novelty, it qualifies as a datum about method; we'll call it the 'virtues datum.'

Recall that a method is a set of criteria that provide instructions for theory construction and measures for theory evaluation. Moreover, a method is sound just in case satisfying its criteria positions inquirers to achieve inquiry's ultimate proper goal—where a method does this just in case satisfaction of its criteria by a theory thereby implies that this theory possesses at least some of that goal's central features and possibly (in nearby worlds) all of them. Given our argument in Chapter 1 (§3) that inquiry's ultimate proper goal is theoretical understanding, these features will include the six characteristics of such understanding discussed there: being accurate, robust, reason-based, illuminating, orderly, and coherent.

The virtues listed above are intimately correlated with these six characteristics. If a method is comprehensive, support-requiring, and synoptic, then it is poised to deliver theories that cover a wide range of data, are supported by positive reasons for belief, and expose systematic relationships in their subject matters. It follows, in turn, that the method is on track to generate theories that have the characteristics of understanding just listed.

If the method is also multidimensional, then it instructs theorists not to privilege one particular criterion, but to take into account a variety of criteria. While possession of this virtue by a theory does not guarantee theoretical understanding, its presence helps position inquirers to achieve it. After all, a theory that is produced by a method that is multidimensional is, if accurate, likelier to answer a broader range of questions, and hence to be more robust and illuminating, than a theory produced by a method that lacks this virtue. Moreover, given that there are six distinct characteristics of theoretical understanding in play, it would make sense that a method poised to deliver such theories will incorporate several distinct criteria.

If the method is also hierarchical and comparative, then it can be used to generate relative—and not just individual—evaluations of rival theories that are sensitive to the diverse qualities of those views. Accordingly, if the method has all of the virtues enumerated above, then it positions inquirers to construct theories that yield theoretical understanding of their subject matters, and to evaluate competing views in light of the extent to which they do this.

3. Extant Methods

To this point, we've identified three data that a sound method must handle, noting that simple methods are ill-equipped to do so.

Here we identify austere formulations of four prominent methods that many philosophers have followed when theorizing. We call these formulations 'austere' because they abstract away from various details and assumptions in order to make explicit each method's core commitments.[14] Although the methods we'll discuss are non-exclusive, each is meant to be complete, in that nothing more than following its instructions is supposed to be required in order to resolve theoretical inquiry. We ourselves think that these methods contain important insights. However, a brief examination of their constituent criteria through the lens of the three data we've collected will allow us to pinpoint challenges each method faces, thereby motivating the quest for a more adequate one, which we seek to provide in the next chapter.

Consider, first, a method that focuses on clarification and instructs theorists to define or analyze a range of central terms, concepts, or properties. We call it the

> **Method of Analysis:** When constructing a theory about a given domain, theorists ought to identify a set of theses about the domain that provide analyses of its central terms, concepts, or properties, where such analyses meet some sufficiently high standard. The best theory is the one whose proposed analyses meet this standard to the highest degree relative to rivals.[15]

The Method of Analysis takes a variety of forms. Some proponents privilege ordinary usage, others formal machinery from contemporary linguistic theory, still others the broadly functional style of analysis known as Ramsification. Yet another option is to

---

[14] The individualism that often structures treatments of method (and which we ourselves reject) is one such assumption.

[15] Proponents or practitioners of this method include Moore (1903), Russell (1914), Stebbing (1932–3), Anscombe (1957), Goldman (1970), Chisholm (1977), Feldman (1992), Strawson (1992), Jackson (1998), Haslanger (2012, Ch. 6), and Hardimon (2017).

privilege the "point" of a term or concept, "the role that [it] plays in our life" or theorizing, and to pursue analyses that potentially refine the meaning or content of the original so as to better realize this purpose.[16] Candidates for the relevant standard include being a universally quantified biconditional, being necessarily true, being knowable a priori, being intensionally correct, being theoretically serviceable, or being conducive to the realization of a certain social-cognitive perspective shift. As foreshadowed above (recall the point about austerity), our characterization omits these details in order to highlight the core commitments of what we take to be a relatively familiar style of theorizing.

A second method, focusing on assembling a view's rationale, is what we'll call the

> **Method of Argument:** When constructing a theory about a given domain, theorists ought to identify a set of theses about the domain that are the conclusions of arguments whose premises and inferences meet some sufficiently high standard. The best theory is the one whose central theses are supported in this way to the highest degree relative to rivals.[17]

---

[16] The quotation is from Craig (1990, 2), who emphasizes practical role and labels his method "conceptual synthesis." Carnap (1950/1962, Ch. 1) privileges theoretical role through an approach he calls "explication." Heidegger (1927/2008, §11) pursues a type of analysis of existence that he calls an "existential analytic." Recently there has been a surge of interest in "ameliorative analysis" (Haslanger 2012, Ch. 6) and "conceptual engineering" (Plunkett 2015 and Cappelen 2018). We believe that at an appropriate level of abstraction all of these approaches can be seen as forms of the Method of Analysis focusing on different standards.

[17] We borrow the label 'method of argument' from Thomson (1990, 29) and Chalmers (2014, 16). Other proponents or practitioners of this method include Harman (1977), Parfit (1984 and 2011), Kamm (1993, 1996, and 2007), Baker (2000), Boghossian (2006), van Inwagen (2006, Lecture 3), Siegel (2010 and 2017), Williamson (2013), Korman (2015), Barnes (2016), Manne (2017), and Khader (2018).

The relevant standard for the inferences will be a broadly logical one: being valid, say, or being cogent, or being such as to provide reason to believe the best explanation (per abduction). The standard for the premises could be: being certain or self-evident, being scientifically or logically well-confirmed, being shared by members of an ideal audience subsequent to extended critical examination, being defensible against relevant challenges, and so on.

A third method focuses on weighing a theory's merits and demerits through what David Armstrong described as "an intellectual cost-benefit analysis."[18] We call it the

> **Cost-Benefit Method**: When constructing a theory about a given domain, theorists ought to identify a set of theses about the domain that possess a high ratio of positive features (benefits) to negative ones (costs), where these features are in themselves roughly equal and aggregative. The best theory is the one that achieves the highest such ratio relative to rivals.[19]

Proponents of this method may select among a range of candidate benefits: rational support, predictive power, explanatory scope, simplicity, conservativeness, and so on. The absence of such features, or possession of contrary ones, would then qualify as costs. The explanation of why a given feature is a benefit could invoke anything from truth-conduciveness, to utility, to aesthetic excellence.

A fourth method, focusing on systematization, is what John Rawls called "wide reflective equilibrium."[20] It has been a mainstay

[18] Armstrong (1989, 19).

[19] Proponents or practitioners of this method include Quine and Ullian (1970, Ch. 6), Lewis (1983 and 1986), Armstrong (1989, §1.4), Gibbard (2003), Weatherson (2003), Godfrey-Smith (2006), Daly (2010, 216–18), Enoch (2011), Paul (2012), Rayo (2013), Sider (2013 and 2020), Nolan (2015, §3), Beebee (2018, §§II and IV), Massimi (2018), and Goldwater (2021).

[20] Rawls (1971, 19–20); *cp*. Rawls (1974 and 1980). See also Goodman (1955, 63–4).

of discussions of philosophical methodology, and is often cited enthusiastically by philosophers wishing to clarify their methodological commitments:

> **Method of Reflective Equilibrium:** When constructing a theory about a given domain, theorists ought to attend to their considered judgments that address the central questions about the domain, seeking to achieve coherence between their considered judgments (at any level of generality) and principles that account for them through a reflective process of modification, addition, and abandonment of either the judgments or principles in case of conflict (with each other, or with any of their other relevant convictions). The best theory is the one that achieves such coherence to the highest degree relative to rivals.[21]

This method is both single-minded, having equilibrium as its sole criterion, and flexible, allowing that modification, addition, and abandonment can occur in whatever way is required to satisfy that criterion.[22]

No doubt there are other methods, but it is fair to say that in contemporary philosophy these four are among the most prominent. And this is no accident, since each of these methods seems poised to deliver at least one proper goal of inquiry. Specifically, it

[21] See the citations in the previous note. Other proponents or practitioners of this method include Daniels (1979 and 1996), Brink (1989, Ch. 5), DePaul (1993 and 1998), Bealer (1996a, 4 and 1996b, 284–6), Elgin (1996 and 2017), Sayre-McCord (1996), Keefe (1999, 42), Pust (2000, §1.4), Scanlon (2002 and 2014), Gozzano (2006), Arras (2007), Lerner and Leslie (2013), Peregrin and Svoboda (2017), Edwards (2018, 20), Finn (2018), and Lycan (2019). Hereafter, we elide 'wide.'

[22] Rawls (1971, 19–20) is responsible for dubbing this criterion 'equilibrium,' which he explicates as "the mutual support of many considerations, of everything fitting together into a coherent whole."

is plausible that the Method of Analysis yields knowledge of definitions, the Method of Argument yields justified beliefs (or perhaps knowledge), the Cost-Benefit Method yields beliefs with a high probability given the evidence, and the Method of Reflective Equilibrium yields a coherent set of judgments. Still, as we'll argue next, it is unclear whether any of these methods puts theorists on the path to realize inquiry's ultimate proper goal.

4. Room for Improvement

We are on the lookout for a method that specifies how to get from the data to a theory that provides understanding, thereby furnishing a satisfactory answer to the question of method. Here we identify reasons to doubt that any of the methods described in the previous section have what it takes to do this.

4.1 Handling the three data about method

One reason is borne of reflection on the data we identified in §§1–2. Handling these three data functions as a test of whether a method is sound.

Regarding the first: the friendliness datum seems to elude the grasp of some of the methods described in the previous section. This datum tells us that, in the paradigm case, implementing a philosophical method involves engaging in such activities as:

- advancing arguments
- raising objections
- offering replies to these objections
- seeking clarification

- seeking explanations
- displaying sensitivity to the deliverances of logic, mathematics, science, and (to a greater or lesser degree) common sense.

Consider the Method of Analysis. Although its constituent criteria comport with an interest in objections, replies, and clarifications, it is not clear that those criteria make sense of the pursuit of arguments and explanations. If all that matters when theorizing is producing a set of theses that (say) identify universally quantified biconditionals that are necessary and knowable a priori, then there is no particular reason for theorists to argue or explain; they could just as easily arrive at these biconditionals by simply consulting intuitions about imaginary cases and collating the verdicts. When it comes to the drive for explanation, the Method of Argument's criteria do not appear to do better. If all that matters when theorizing is assembling the rationale for one's view, then hunting for explanations is beside the point.

Turn now to the determinativeness datum, according to which a sound method selects among multiple theories that fit the data, ruling a great many of them out of contention. We're unsure whether any of the four methods can handle that datum. This for two reasons. First, all four methods are silent about data; there is nothing in their constituent criteria that directs theorists to assemble claims that handle them.[23] Second, even if the methods were supplemented in order to do this, questions about theory selection

---

[23] That our formulations make no mention of data flows from our aspiration to fidelity, to capture the way that these methods are standardly cast in the literature. But each method arguably recognizes inputs, in the form (say) of premises or considered judgments, that could be cast as data. Whether this results in an acceptable treatment of data is an important question, but one we won't pursue here.

would remain. *Perhaps* each method's criteria could be outfitted in such a way as to guide inquirers to a small set of theories.[24] But the devil is in the details.

This brings us to the virtues datum, which says that a sound method is comprehensive, support-requiring, synoptic, multidimensional, hierarchical, and comparative. In our estimation, each of the four methods fails to possess the entire suite, rendering each either incomplete (lacking at least one virtue) or vicious (obstructing at least one).

At their best, the Methods of Analysis and Argument may be comprehensive, support-requiring, multidimensional, and comparative. But there looks to be nothing in either method's criteria that makes it particularly likely to be synoptic or hierarchical. Even when embellished with sophisticated standards and applied iteratively, the fact remains that these methods' constituent criteria do not instruct theorists to identify systematic relationships, including explanatory ones, of the sort implied by the first feature. Nor does either method contain instructions for ranking its criteria in the way required by the second feature. Consequently, both methods are incomplete.

As for the Cost-Benefit Method, it is multidimensional and comparative, for it sanctions a plurality of criteria, such as rational support, explanatory power, coherence, and simplicity; it then treats them as roughly equal and aggregative, such that the comparative merits of various theories are determined by simply

[24] The Method of Reflective Equilibrium may be an outlier. Supposing we construe the data set as the tranche of considered judgments that initiate the reflective process, the method instructs inquirers to assemble theories exhibiting a high degree of coherence with that set. All such theories will fit the data. But the method fails to whittle down these candidates, thus contravening the determinativeness datum.

weighing them up. Above we described this using the economic idiom favored by Armstrong and others, such as Lewis, who counsels theorists to "measure the price."[25] David Enoch has more recently employed the language of "tallying plausibility points," acquired by the number and quality of answers proffered in reply to a set of questions. Answering a question earns a theory some points; that the answer is explanatory earns it some more; that the answer is coherent or parsimonious earns it yet more; the absence of one or more of these things means the loss of points; and so forth. Tally all the points accrued, measuring the "gains" against the "losses," and whichever theory has the best score—that is, "the most plausibility points"—wins.[26]

The upshot is that the method permits theorists to fail to answer a range of questions, so long as it does extremely well in other respects. This makes it insufficiently comprehensive. The method also allows for theories that lack rational support or explanatory power, thus rendering it insufficiently support-requiring and synoptic. Because the method treats all positive features as on a par, any such feature can be sacrificed, leaving the method insufficiently hierarchical. In these ways, it does not handle the virtues datum.

At the root of the trouble is the Cost-Benefit Method's tolerance for illicit trade-offs, stemming from the method's signature commitment to treating "plausibility points" as roughly equal and aggregative. When it comes to understanding, it is far from clear that a gain with respect to one criterion compensates for losses with respect to others. For instance, if a theory achieves a high level of parsimony, but does so at the expense of explanatory

---

[25] Lewis (1983, x).   [26] Enoch (2011, 267).

scope and rational support, then this trade severely diminishes the theory's ability to promote understanding. If so, then its parsimony may not win it *any* points, much less any that could compensate for shortcomings elsewhere. That the Cost-Benefit Method condones such questionable behavior marks it as not just incomplete but also potentially vicious.

Turn next to the Method of Reflective Equilibrium. This method clearly possesses two of the methodological virtues, being both comprehensive and comparative. It may also be synoptic. Unfortunately, however, the method is incomplete, and potentially vicious.[27]

To see why, recall that it permits theorists to introduce new claims of various sorts, so long as doing so increases equilibrium. And in fact theorists will be forced to make such additions if they are to avoid simply reprising the method's starting points (viz., the initial considered judgments). The method thus opens the door to systematic theories whose claims are unsupported by any consideration, beyond coherence, that speaks in their favor. This raises the worry that the Method of Reflective Equilibrium is not adequately support-requiring. It sanctions outputs that belong "in a box with rumors and hoaxes," to quote Wilfred Sellars' barb targeting a narrow focus on coherence.[28]

In addition, the method looks to be insufficiently multidimensional and hierarchical, since (like the Cost-Benefit Method) it

---

[27] There have been many critics of a method focused on reflective equilibrium. In a number of cases, the concerns they voice are plausibly interpreted as leveling this charge. See, e.g., Hare (1973), Brandt (1979, 19–21 and 1990), Raz (1982), Copp (1985), Stich (1988), Cummins (1998), Bonevac (2004), Kelly and McGrath (2010), McPherson (2015 and 2020, §5.3), and McGrath (2019, Ch. 2).

[28] Sellars (1956, §38).

appears to license illicit trade-offs. It is possible that equilibrium is sometimes achievable only by forgoing extensionally adequate or rationally supported principles for those that lack these features. As one possibility, imagine that an accurate claim regarding some domain is extremely complicated; as such, it is in tension with various other commitments. Imagine further that an alternative, inaccurate claim about the domain is much less complicated, and so does not induce the same tension. In that case, it's possible that we greatly enhance equilibrium through the second claim. Even if accuracy may sometimes be exchanged for other features, the sort of trade just sketched strikes us as a bad deal.

If these concerns are legitimate, then the Method of Reflective Equilibrium fails to handle the virtues datum. Since the method omits and perhaps obstructs one or more of the rudiments of a virtuous methodology, it is at best incomplete, and at worst vicious.

4.2 Delivering inquiry's ultimate proper goal

To this point we've submitted a negative evaluation of the four methods in light of their poor showing with respect to the three data about method. There's a second reason for that evaluation, one that emerges from a direct appraisal of the capacity of those methods to deliver inquiry's ultimate proper goal. We've defined a sound method as one such that successfully following its instructions positions inquirers to achieve this goal, which we take to be theoretical understanding. Such understanding requires full grasp of a theory that is accurate, reason-based, robust, illuminating, orderly, and coherent. We have no compelling reason to think that any of the methods considered above is on track to yield theories possessed of all six features, and some reason to think that each is bound to fall short.

The Method of Analysis, even when generously interpreted so that its implementation yields a set of analyses that are accurate, reason-based, coherent, and offer some explanatory illumination of the analysanda, may nevertheless fail to yield theories that do all the explaining that needs doing. Nor is the method poised to deliver theories that are robust, answering a domain's central questions (which are many and diverse), or orderly, transcending a set of feature-specific explanations to afford a broader view of how the various elements of the domain hang together. The Method of Argument is in a similar boat. For it simply ratifies conclusions of arguments, which (as we noted in the Introduction) fall far short of theories, let alone ones that are robust, orderly, and illuminating. Inquirers can analyze and argue til the cows come home, but analyses and conclusions must be integrated and explained in order to deliver a theory that yields understanding of a domain. Such integrative and explanatory work is beyond the purview of the Methods of Analysis and Argument.

Our concerns about the other two methods take a different shape. With the Cost-Benefit Method, there is a sense in which everything is up for grabs. It does not select theories that at once possess all six features of understanding. For example, if a theory manages to be highly illuminating, robust, orderly, and coherent at the expense of accuracy, then the method licenses treating the last as expendable. In another context, a different array of costs and benefits could make different features expendable. By contrast, successfully implementing the Method of Reflective Equilibrium guarantees that a theory will at least be robust, orderly, and coherent. But the method infamously fails to put inquirers on track to achieve even a modicum of accuracy; nor is it clear that it plumps for outputs that are reason-based in the indicated sense. Indeed, if the modifications, additions, or abandonments that help inquirers achieve equilibrium are wide of the mark, lack support, or undermine robustness, then the equilibrium attained will fail to promote theoretical understanding.

To be clear, in rendering these verdicts we're not claiming that all four methods are useless. As we've noted, these methods may help inquirers achieve a proper goal of inquiry, such as justified belief or knowledge.[29] However, the question at hand is whether there is a method that is sound, poised to deliver a theory that realizes inquiry's *ultimate* proper goal, namely, theoretical understanding. Considered in this light, none of the methods we've canvassed meets the bar.

4.3 Remedies?

Advocates of these methods are free to attempt to remedy the deficiencies we've identified so that their preferred method fully handles all three of the data. But it's unclear whether these methods are amenable to repair. For such revisions would hardly be incidental refinements; rather, they'd amount to substantial alterations to the method's basic instructions.

For example, consider the Method of Reflective Equilibrium. In order to ensure that this method is sufficiently hierarchical, its instructions would need to be revised so as to prohibit the possi-

---

[29] Notably, Rawls (1980, 534) described reflective equilibrium as merely uncovering the doctrine that is "most reasonable for us" to accept; similarly, Goodman (1955, 61–2) and Brink (1989, 140–1) emphasize that reflective equilibrium provides justification, allowing that it does not facilitate other, stronger epistemic achievements.

bility of attaining equilibrium through bad deals (as described above). That, however, appears to introduce criteria that privilege something other than equilibrium—a revision of the spirit, and not simply the letter, of the method. Or to ensure that the Cost-Benefit Method does not license illicit trade-offs, proponents would need to modify its constituent criteria in such a way as to deny that the various benefits (and corresponding costs) are roughly equal and aggregative. Our point is not that this revision is impossible, but that it risks turning the method into something it is not.

To further appreciate the challenge, consider a few possible responses that defenders of these methods might offer. Speaking on behalf of the Cost-Benefit Method, Chris Daly writes that

would-be critics of the method may not so much reject the method as reject the weighting given to various costs and benefits....The locus of the debate...will then not be over the appropriateness of the cost-benefit method. It will be over which factors are costs, which are benefits, and what weightings to give them.[30]

This point serves as an apt reply to certain sorts of charges levied against the Cost-Benefit Method. However, our critiques do not rely on acceptance or rejection of any particular weighting scheme. Rather, they target the method's commitment to the basic idea that various benefits—whatever they might be—are roughly equal and aggregative, such that they can be allied with and pitted against one another so as to license particular sorts of trade-offs. While some may promote understanding, others do not, and so a sound method must include criteria that ban them.

[30] Daly (2010, 217–18).

Proponents of the Method of Reflective Equilibrium have two replies of their own to the charges we've raised. The first appeals to a certain interpretation of this method, one that emphasizes its flexibility. Kenneth Walden, for example, recommends an "anti-essentialist" view whose signature implication is that "the method of reflective equilibrium is not, exactly, *anything*."[31] That is,

The method of reflective equilibrium should be understood not as a particular method that lays out theses about the nature of [data], methodology, and criteria for epistemic success, but as what is left over of our methods when we deny that we can do this.[32]

Such a position may well serve as a "tonic corrective" to the "temptation to take some…contingent feature of inquiry and cast it as the very essence of inquiry."[33] However, viewed as a reply to the worry we've aired, such a position is inadequate for at least two reasons. First, it does not provide any means for securing friendliness, determinativeness, or the virtues that the method lacks. Second, it overreaches. To the extent that the point and purpose of method is to guide and constrain theorizing, the sort of vacuity countenanced by this approach must be regarded not as a defense of a specific method, but rather as a retreat to *skepticism* about method.

The second response we'll consider adverts to the idea that the Method of Reflective Equilibrium is the only game in town. Thomas Scanlon writes:

[31] Walden (2013, 244).   [32] Walden (2013, 255).
[33] Walden (2013, 255). Derrida, a general critic of essentialism, famously made similar remarks about his preferred method, deconstruction, eventually writing: "What is deconstruction? Nothing, of course!" (1991, 275).

[I]t seems to me that this method, properly understood, is in fact the best way of making up one's mind about moral matters and about many other subjects. Indeed, it is the only defensible method; apparent alternatives to it are illusory.[34]

The last claim of this passage strikes us as incorrect. Think of the different methods we've examined so far. They privilege conceptual analysis, argument, or miscellaneous benefits (such as parsimony or predictive power). In the next chapter, we'll detail a method that privileges accommodation, explanation, substantiation, and integration. None of these criteria is equivalent to equilibrium. Consequently, if the assertion that there is no alternative to the Method of Reflective Equilibrium is to stand a chance of being true, it can be so only because it supposes that this method need not prioritize equilibrium, but can sometimes give most weight to other criteria. That is not, however, an adequate defense of the method, and not simply because it risks draining the method of its distinctive content. For the assertion itself does not support the claim that appealing to reflective equilibrium satisfactorily answers the question of method.

What about a method that conjoins the four methods we've described—might such an "all-inclusive" method evade our objections? Perhaps. But we are not sanguine about the prospect. The four methods appear to privilege criteria that point in very different directions, making it unclear whether and how they can be consistently conjoined. It seems to us that integration could be achieved only at the cost of introducing revisions that substan-

---

[34] Scanlon (2002, 149). *Cp.* DePaul (1998), as well as Keefe's (1999, 42) comment: "There is, I suggest, no possible alternative methodology."

tially alter the letter, and violate the spirit, of at least some of these methods.

If our critical remarks have been on target, the question of method—how to get from data to a theory that realizes inquiry's ultimate proper goal?—is as yet unanswered. In the next chapter, we propose our own solution.

5

The Tri-Level Method

A sound method positions inquirers to achieve theoretical understanding. This chapter presents a method designed to do just that. We endorse the method not because it makes a philosopher's job easy; indeed, it is quite demanding. Nor are we drawn to its constituent criteria because they revolutionize philosophical thinking; on the contrary, all of them are familiar from the way many philosophers go about their business. However, they have not yet been combined, integrated, and clarified in a way that reveals how their conjunction can promote inquiry's ultimate proper goal: theoretical understanding. Because these criteria operate at three distinct levels, we call the method they compose the 'Tri-Level Method.'

Its first level directs theorists to handle the data in a given domain through a series of claims and commitments. The second instructs theorists to ensure that these claims and commitments, which comprise their theories, are themselves well-grounded,

Philosophical Methodology: From Data to Theory. John Bengson, Terence Cuneo, and Russ Shafer-Landau, Oxford University Press. © John Bengson, Terence Cuneo, and Russ Shafer-Landau 2022. DOI: 10.1093/oso/9780192862464.003.0006

epistemically and explanatorily. The third charges theorists with the task of rendering their theories virtuous in certain respects. These are the building blocks of the

> **Tri-Level Method**: When constructing a theory of a given domain, theorists ought to articulate a set of theses about the domain that (i) accommodate and explain the data, (ii) are themselves substantiated and integrated, and (iii) possess specific theoretical virtues. All of the criteria at the first two levels, given by (i) and (ii), include escape clauses. Each level takes priority over its successors. The best theory is the one that satisfies the criteria at these levels (so ordered) to the highest degree relative to its rivals.

This austere formulation of the Tri-Level Method hints at how the method preserves insights in the four popular methods from Chapter 4. It does so partly through the assignment of important roles to analysis (which may subserve accommodation, explanation, or substantiation), argument (as an element of substantiation), equilibrium (since coherence is central to integration), and miscellaneous theoretical benefits (in the form of virtues). Yet the Tri-Level Method neither reprises those methods nor simply conjoins their disparate criteria. The method calls for much more than mere analysis, argument, equilibrium, or theoretical virtues, while also rejecting the notion that its criteria are roughly equal and aggregative. This sets it apart.

We offer the following diagram of the method's five criteria, which visually depicts their organization.

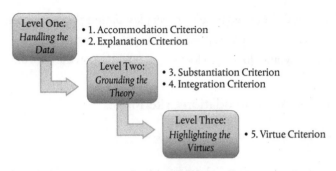

The Tri-Level Method

Levels one and two both feature a pair of criteria. The pair at the first level enjoys priority over the pair at the second (in a sense we'll explain below). All four criteria at the first two levels enjoy priority over the sole criterion at the third level, which (on our version of the method) plays only a tie-breaking role between theories that do well enough at lower levels.

Both the austere formulation and the diagram above showcase the basic content and organization of the Tri-Level Method's constituent elements. While our official statements of the method's five criteria will aspire to neutrality, we'll also fill in many of the specifics, albeit in ways that allow for further elaboration and refinement. The primary aim of this chapter, then, is not to hammer out every last detail, but to present the core elements of the Tri-Level Method and their arrangement with an eye to how they work together in the service of understanding.

1. Level One: Handling the Data

Data are the inputs to theorizing. The function of the two criteria at our method's first level is to ensure that theories handle the data.

Theorists do not need to collect every last datum about a domain in order to succeed here. Rather, as anticipated earlier (Chapter 3, §2), theorists would do well to focus their attention on the core data.

1.1 The Accommodation Criterion

The first instruction is given by the

> **Accommodation Criterion**: A theory of a given domain must accommodate the data in that domain, or at least give an adequate defense of the claim that those data require no such accommodation,

where a theory T 'accommodates' a datum just in case that datum is likely to hold or be true, given T. We offer three clarificatory remarks regarding this criterion.

First, while we leave open the precise analysis of the relevant notion of likelihood, its basic meaning and application are intuitive. Recall, for example, the datum that aesthetic judgments are intimately connected to personal preferences. While both realist and antirealist views of the aesthetic domain are logically consistent with this datum, the latter view's assertion that aesthetic features depend on our attitudes clearly renders this datum likely in a way that the former does not. This is a theoretically important difference between the two views, which the Accommodation Criterion recognizes.[1]

Second, the Accommodation Criterion incorporates what is, in effect, an escape clause: rather than accommodate a given datum,

---

[1] Any adequate analysis of the relevant notion of likelihood will preserve at least three important properties: first, it is stronger than logical consistency (see the example in the text); second, it allows that a datum may be likely on a given theory T, even though T does not explain that datum (an example is given in §1.3); and, third, whether a datum is likely, given T, is not fully determined by that datum's modal status.

a theory can instead adequately defend the claim that this is not required. Such a defense, in our judgment, can take one of two forms. It might involve providing a good reason[2] to believe that the theory coheres well with an independent set of claims that fully accommodates that datum (in which case the theory need not do this itself). Or it might involve identifying a data disabler, which supplies an undefeated epistemic reason that defeats the reason favoring the datum in question (as discussed in Chapter 3, §3).

Third, a view might satisfy the Accommodation Criterion to a greater or lesser degree. There are both quantitative and qualitative dimensions to this last point. Quantitatively, a theory might meet this criterion with respect to some but not all of the data; it requires supplementation in order to handle the remainders. In such a case, that theory incurs burdens of accommodation that can be discharged only by augmenting the original view. If the amended view manages to discharge those burdens, then it fares better than the original by the lights of this criterion. Qualitatively, a theory might do a stellar job at satisfying the criterion with respect to a data set, perhaps by rendering its members highly likely. Alternatively, a theory might render them just barely likely, in which case the theory would satisfy the Accommodation Criterion, but only passably.

1.2 The Explanation Criterion

It is not enough for a theory to render the data likely. Its job is also to *explain* them. This is the point of the

---

[2] In Chapter 3 (§1), we characterized a good reason as an undefeated epistemic reason for belief that would not be easily defeated by competing considerations, and whose possession implies the possession of at least some evidence.

Explanation Criterion: A theory of a given domain must explain the data in that domain, or at least give an adequate defense of the claim that those data require no such explanation,

where a theory T 'explains' φ just in case T invokes some ψ such that φ holds or is true because ψ holds or is true. As above, we make three remarks about this criterion.

First, we leave open the precise analysis of the notion of one entity's holding or being true *because* another entity holds or is true. Presumably, a satisfactory analysis will countenance diverse forms of explanation (e.g., causal, unificationist, modal, constitutive, grounding, essentialist), each of which implies potentially domain-specific conditions of adequacy.[3] As for the epistemology of explanation, it can be perfectly appropriate for an explanans to have a weaker epistemic standing than the datum it explains—after all, data are often backed by excellent reasons, and it would be a tall order to always marshal an explanans with similar credentials.

Second, the defense cited in the Explanation Criterion's escape clause can take one of three forms: it might involve (i) providing good reason to believe that the theory coheres well with an independent set of claims that themselves explain the data (in which case the view need not do this itself, as above); (ii) identifying a data disabler; or (iii) supplying good reason to believe that the datum does not admit of explanation, but is instead explanatorily basic.

Third, as in the case of the previous criterion, a theory might satisfy the Explanation Criterion to a greater or lesser degree, both quantitatively and qualitatively. Quantitatively, in some cases, a view will explain most but not all of the data—the remainders

---

[3] Throughout we elide 'adequate,' speaking simply of 'explanation' and 'explaining.'

qualify as burdens of explanation. Should the view be supplemented so as to discharge some of those burdens, then the resulting view has a leg up on the original. Qualitatively, a theory might deliver a deep explanation, or instead provide a minimally adequate explanation, of a datum. The former is obviously better.

1.3 On the relation between the first two criteria

While the Accommodation and Explanation Criteria both focus on the data, they mark two different, though complementary, intellectual projects: the former criterion issues a defeasible norm for theorists to certify the likely truth of a datum, whereas the latter issues a defeasible norm for them to explain why it is true.

These criteria also potentially diverge, at least insofar as a theory may render a datum likely without explaining it. This is, for example, one way of interpreting the charge that nonnaturalist moral realism fails to account for the relation between moral and non-moral features, as codified by various moral supervenience theses: the view implies the likelihood of the supervenience of the moral on the non-moral, yet fails to explain it (or certain of its features). The idea that a theory explains a datum without accommodating it is a bit less straightforward. Arguably, in many and perhaps all cases, if a view explains a datum, then that datum is also going to be likely, given that view. Perhaps there are exceptions in which explanation does not entail accommodation.[4] But as this will not affect anything in what follows, we can afford to remain neutral about this matter here.

Satisfying the Explanation Criterion without remainder is difficult. In some cases, it may be a significant achievement to explain

---

[4] *Cp.* Clatterbuck (2020).

even a limited range of the data. When there are many remainders, however, the view is likely to be deficient. By contrast, if the remainders are relatively few, the theory may be in good shape, poised to provide theoretical understanding of its subject matter.

Indeed, both criteria at level one target features of understanding. By satisfying the Accommodation Criterion, a view has a greater chance of being *accurate*; by fulfilling the Explanation Criterion, it offers a greater promise of being *illuminating*; by jointly doing these things with respect to a wide range of data, the theory achieves *robustness*.[5]

Given the importance of these three features to understanding, and their relation to the Accommodation and Explanation Criteria, level one occupies a special place in the Tri-Level Method. This is captured in part by the following condition of adequacy: a theory is minimally adequate, standing a chance of facilitating understanding of the domain, only if it does modestly well with respect to both criteria at level one.[6] In the pursuit of understanding, greater success is wanted, but at least this much is needed.

2. Level Two: Grounding the Theory

The criteria at the first level of the Tri-Level Method select for theories that handle the data. But theorizing cannot stop here. For, in

---

[5] How do the two criteria at level one square with models, which often diverge from what they represent? We allow that models can play various roles in theorizing, including accommodation and explanation of data. Models can do these things so long as they are accurate or are supplemented with the claim that their divergence from what they represent is negligible for relevant cognitive purposes—a claim that preserves the theory's accuracy (Bengson 2020, §3). Similar points may apply to idealizations, metaphors, analogies, and narratives.

[6] We can afford to remain neutral on what it is for a theory to do 'modestly well,' though we are inclined to think that it excludes wholesale use of the escape clauses.

handling the data, theories make claims or have commitments that raise questions about their epistemic and explanatory credentials, as well as about their relation to scientific theories and common sense; it is vital that theorists forthrightly tackle these questions. This brings us to the second level of the Tri-Level Method, which is designed to ensure that the theory's claims and commitments are well-grounded.

2.1 The Substantiation Criterion

At the second level, a theory is first and foremost called upon to defend and explain—in short, to 'substantiate'—its claims (i.e., constitutive theses) and commitments (i.e., implications of those theses and their applications). While we've already characterized the notion of explanation with which we'll work, we should say more about the notion of a defense.

We distinguish two types of defense. The primary type consists in the provision of positive epistemic support for a given claim. The relevant type of support, in our view, consists in epistemic reason for belief. Such a reason needn't be dialectically persuasive, capable of rationally convincing an opponent. Neither must it be indefeasible, or even conclusive. Still, it needs to be good (both strong and undefeated).[7] A secondary type of defense consists in responding to genuine objections to one's views. Here, theorists are "playing defense"—responding to extant or anticipated criticisms, rather than building a positive case on behalf of their view.

---

[7] We leave various details open. For example, though we'll often focus on reasons for belief, we are officially neutral on the relevant attitude (whether it is a type of endorsement or something else, such as acceptance, as suggested by critics of the idea that belief is the appropriate attitude to take towards philosophical theses).

By diagnosing and adequately replying to objections, this sort of defense can buttress the positive case advanced in a primary defense, strengthening the support for the theory's claims and commitments. But as this suggests, secondary defenses ride piggyback on primary ones: the former are not reasons for belief in the view, but at most bolster the reasons provided by the latter.

We dub the call for defense and explanation of a theory's claims and commitments the

> **Substantiation Criterion**: A theory of a given domain must substantiate its claims and commitments, or at least give an adequate defense of the claim that it is not required to do so.

We make four remarks about this criterion before identifying its primary rationale.

First, the Substantiation Criterion applies to any claim or commitment made by a theory. Among these are ones it makes at the first level, when handling the data through satisfaction of the Accommodation and Explanation Criteria. But these are not all. The Substantiation Criterion also applies to all further claims and commitments enlisted by a theory at levels two and three.

Second, satisfaction of this criterion's escape clause can take one of two forms. It might involve identifying good reason to think that a theory coheres well with an independent set of theses that substantiate the claim or commitment in question (in which case, once more, the theory need not do this itself). Alternatively, it might involve providing good reason to believe that a claim or commitment requires no substantiation. While we're skeptical that theorists can evade the call for defense, they can avail themselves of the escape clause when it comes to explanation by citing a good reason to believe that a given claim or commitment is an unexplained explainer. Any unfinished business—claims or commitments that have not met the demand set forth in this criterion—qualify as burdens of substantiation.

Third, although we leave open the precise analysis of various aspects of the Substantiation Criterion, we recognize the following constraint: a theory's explanations of its claims or commitments must not invoke a datum that they themselves explain, on pain of circularity. However, we allow that a view's defenses of its claims and commitments may invoke a datum that they themselves handle. They might even invoke the fact that they handle it. (Such reasoning may, for instance, be part of an abductive defense.) Note, though, that in any such scenario it would be neither the data nor the fact that they are being handled that alone provided the needed defense; additional claims, in all likelihood substantive ones (e.g., that the claim or commitment in question provides the *best* explanation of a datum), would be needed.

Fourth, the Substantiation Criterion is similar to the criteria at level one in that it may be satisfied to a greater or lesser extent. As above, this point has two dimensions. Qualitatively, defenses and explanations can be better or worse, depending on the strength of the reasons they provide or the degree of illumination they offer. Quantitatively, we recognize that a theory of a given domain will ideally go on substantiating its claims and commitments until every single one that admits of defense and explanation is fully defended and explained, and every defensible and explicable claim or commitment enlisted in that enterprise is itself fully defended and explained, and so on *ad indefinitum*. However, at some point the claims and commitments may be so far removed from the target domain that defending and explaining them will not substantially further understanding of that domain. A theory that fails to substantiate such claims and commitments would not be ideal, though it may still achieve excellence.

The rationale for the Substantiation Criterion is straightforward: a theory affords greater understanding of its target to the extent that it substantiates its claims and commitments. By defending its claims and commitments, a view becomes *reason-based*; by explaining its claims and commitments, it adds *robustness* and overall *illumination*. Doing these things together enhances the prospect of greater *accuracy*. Consider a view that accommodates and explains all of a domain's data only by invoking a set of undefended and unexplained claims and commitments. When forced to choose between such a view and an alternative theory that accommodates and explains the same data through claims and commitments that receive defense and explanation (substantiation), we clearly ought to pick the latter, which will possess those four features of understanding to a much greater extent.

2.2 The Integration Criterion

We turn now to the other criterion embedded in our method's second level, which we call the

> **Integration Criterion:** A theory of a given domain must ensure that its claims and commitments integrate with each other and our best picture of the world, or at least give an adequate defense of the claim that the absence of such integration is unproblematic,

where a theory T 'integrates' with φ to, and only to, the extent that T is not just consistent with but coheres with φ. This characterization exposes one of two rationales for demanding integration: in its absence, a view lacks the *coherence* displayed by theories that yield understanding. The second rationale is that on the assumption that our best picture of the world is in good shape, integration increases the prospect of *accuracy*.

This assumption is fairly safe given that our best picture of the world is to be understood in a broad and relatively non-committal way as designating a range of claims supported by good reasons—specifically, those uncontroversial claims furnished by common sense and our best physics, biology, chemistry, medicine, logic, and mathematics.

Unsurprisingly, it is a complex matter to specify, at any given time, what is furnished by common sense. For present purposes, let's say that a claim is so furnished only if it is accepted by both you and us; additionally, it is a sufficient condition that the claim be common ground between your family and our families. That lightning tends to signal rain, that tigers have stripes, that being hit

Illustration 6 Panthera tigris.

often hurts, that most human beings know their own names—these make the cut. That a particular philosophical view such as libertarianism or hard determinism is true, or that a specific philosophical thesis such as the doctrine of double effect is correct, does not.[8]

Some philosophical views contain conflicts, thereby lacking internal integration. Others conflict with what is furnished by our best picture of the world, and so lack external integration. Whether internal or external, such conflicts either do or don't involve strict incompatibility. When they don't, theorists may add a claim that resolves the tension (e.g., by forging connections between apparently discordant claims). When they do, theorists must provide good reason for denying that such conflict is problematic—per the Integration Criterion's escape clause. Either strategy would relieve a theory of a burden of integration, which would otherwise remain.

2.3 On the relation between the first two levels

The Substantiation and Integration Criteria identify two distinct tasks. But they are related in that both go beyond the criteria at level one (accommodation and explanation of the data) by issuing defeasible norms for theorists to ensure that the claims and

---

[8] We've (1) offered a general characterization of the notion of our best picture of the world, (2) identified a substantive necessary condition for inclusion in our common-sensical conception, and (3) proposed a substantive sufficient condition for such inclusion. This triangulation and the accompanying examples provide an informative initial explication of the notion of our best picture of the world, while respecting the prospect of further elaboration and refinement.

commitments of the theories are themselves well-grounded. Theorists may proceed serially, tackling the criteria at level one before turning to those at level two. But they need not. It often makes sense to do some work at level two before trying to complete the job at level one. For example, the explanans of a thesis introduced to defend a claim that accommodates a datum might do double-duty as an explanation of that datum.

The two levels are complementary in at least one further respect. When the Substantiation and Integration Criteria, along with the Accommodation and Explanation Criteria, are each fulfilled to a high degree, the resulting view is not only on track to be accurate, illuminating, robust, reason-based, and coherent, but also poised to achieve the *orderliness* that is characteristic of theories that promote understanding of their subject matter.

In §1.3 we said that a theory is minimally adequate if and only if it does modestly well with respect to both of level one's criteria. Suppose that a minimally adequate theory also manages to do modestly well with respect to the criteria at level two. Then, and only then, is the theory 'respectable,' being on track to deliver a modest level of understanding. This means that the two levels are both important, though not equally so. Yet the priority of the first level is defeasible in the sense that extensive successes at level two may deliver substantial improvements in understanding that are only barely diminished by small failures at level one. (The reverse is also true.)

3. Level Three: Highlighting the Virtues

Most philosophers have expressed a preference for views that are more virtuous—for example, more modest, parsimonious, natural, fruitful, or beautiful—than their rivals. We agree that virtues matter under certain conditions. The key question is what these conditions are. Suppose that all is more or less equal among multiple theories at the Tri-Level Method's first two levels. In such conditions, and only in such conditions, theorists are called on to ascend to the method's third level, which instructs them to attend to the virtues of a theory as a whole. Such are the topic of the

> **Virtue Criterion**: All else being equal, a theory of a given domain must be more theoretically virtuous than rival theories.

After making two points about this criterion, we'll address its motivation.

First, it bears emphasizing that this criterion targets the virtuosity of a *theory*, and thus concerns the *being-theoretically-more-virtuous-than* relation when and only when its relata are whole theories. For all this criterion says, appeals to simplicity and various other putative virtues might play different roles when what is at issue is, say, the adequacy of a particular explanation advanced by a theory. In that case, we might note that one explanation is more virtuous than another, and so is more adequate. That might be directly pertinent when considering the call for explanation issued by the Explanation and Substantiation Criteria. But it would bear upon the wholesale evaluation of the theory only indirectly, if at all, since it would not by itself imply that a theory that incorporates this explanation is virtuous, or more virtuous than rivals.

Second, we interpret the prefatory clause 'all else being equal' to signify that the Virtue Criterion plays just a tie-breaking role, and does so only when two conditions are in place: (i) the candidate theories are all respectable, doing at least modestly well at levels one and two; and (ii) the candidates are roughly equal—that is, more or less on a par—with respect to the four criteria at those levels. This precisification is not mandated by the Tri-Level Method itself. But for reasons offered below, we find it compelling. Since levels one and two will often suffice to identify one theory as better than its rivals, there will be a wide range of cases (most, we suspect) in which this fifth criterion has no role to play whatsoever. Indeed, in practice and in principle, the second level of the Tri-Level Method may mark the terminus of theorists' efforts. For it may be that one theory of a given domain does a superior job accommodating and explaining the domain's data, and substantiating and integrating its claims and commitments.

There is a long tradition, running from Aristotle through Aquinas, Occam, and Kant, that endorses the Virtue Criterion, complete with its ceteris paribus clause. Some of these philosophers have maintained that theoretical virtues play a role in theory choice because they've assumed that simplicity and other candidate virtues are truth-conducive: *simplex sigillum veri*. We do not help ourselves to this assumption. Rather, because our method is designed with an eye toward promoting understanding, it ratifies a candidate virtue as genuine if, but only if, its possession somehow contributes to or enhances understanding. Any alleged virtues that do not meet this bar fail to qualify as theoretical virtues, and thus fail to be relevant to the Virtue Criterion. We don't take a stand here on which virtues pass muster.

Our perspective on the role of theoretical virtues follows from the conjunction of two claims:

(i) None of the theoretical virtues, either by itself or in tandem with others, is such that its possession contributes to or enhances the view's capacity to provide understanding of its subject matter.

(ii) Once the Accommodation, Explanation, Substantiation, and Integration Criteria are satisfied at least modestly well, then a view's exemplifying a theoretical virtue *does* contribute to or enhance the view's capacity to provide understanding of its subject matter.

The conjunction can be motivated by the following sorts of considerations.

Consider the virtue of simplicity as a test case for claim (i). The solipsist position introduced by Descartes toward the beginning of his *Meditations* is strikingly simple. This position states that there exists exactly one person—you—and your mental states. Solipsism is, on at least one construal of simplicity, a much simpler theory than one that posits you and your mental states *plus* an external world, distinct from you, inhabited by some untold number of types of entities and tokens of those types. And yet solipsism is a crazy view. It is crazy largely because it so thoroughly disregards the data that any view regarding what there is must accommodate and explain (including the existence of your fellow human beings), fails to defend and explain its own claims and commitments (including how there can be you and only you), and flouts the demand to attain integration with our best picture of the world. Nor does it exploit the escape clauses by providing adequate defenses that free theorists from having to do these things. *Perhaps* solipsism can account for these data, substantiate its claims and commitments, and achieve internal and external integration, or satisfy all of the relevant escape clauses. The point is that the bare fact that solipsism is simpler than the external-world view does not, in and of itself, promote theoretical understanding; the relative simplicity of solipsism does not by itself speak in its favor.[9]

This argument for claim (i) can be extended to other putative theoretical virtues. It implies that while such virtues may play an important methodological role, as many have supposed, their significance is more restricted, and more modest, than is sometimes assumed. They do not by themselves increase the chance of securing any of the four fundamental features of a theory that provide understanding: being accurate, illuminating, reason-based, or robust. Granted, some virtues may help to secure the other pair of features, coherence and orderliness, whose contributions to understanding are conditional (as noted in Chapter 1, §3). Or they may otherwise boost the contributions to understanding made by accommodation, explanation, substantiation, and integration. We submit that the job of such virtues is plausibly construed in the manner described in claim (ii), which implies that their role is restricted to that of being tie-breakers among respectable theories.[10]

We recognize that some will see things differently, contending that the methodological role of theoretical virtues is more powerful than this construal allows. For example, it might be said that,

---

[9] That it is also (say) general and elegant does not improve its prospects. The reasoning in the text is of a piece with our remarks about parsimony in Chapter 4 (§§2.1 and 4). See Huemer (2009), Sober (2009), and Kriegel (2013, §3) for critical discussion of the role of certain putative theoretical virtues in philosophy.

[10] We are sympathetic, but not wedded, to Zagzebski's (2001, 244) suggestion that in some cases "understanding is achieved partly by simplifying what is understood, highlighting certain features and ignoring others." Theoretical virtues may do this by, for example, making a respectable theory more orderly—one of the six features of an understanding-providing theory.

provided that a theory is respectable, its virtues tip the scales in its favor if it is only moderately inferior to competing theories at levels one and two.

Strictly speaking, this proposal is compatible with the Tri-Level Method. But it is not our view. For the proposal doesn't fully respect the difference that has emerged between the sole criterion at level three and the four criteria at levels one and two. We've argued that the Accommodation, Explanation, Substantiation, and Integration Criteria, but not the Virtue Criterion, are constitutively linked to the four fundamental properties of an understanding-providing theory: being accurate, illuminating, reason-based, and robust. If a theory fares poorly in comparison to its rivals in light of the criteria at levels one and two, then it is *less* likely than its rivals to possess these properties to the highest degree. It follows that the theory is not the one that is most likely to furnish understanding—even if it far surpasses its competitors with respect to simplicity, or is more theoretically virtuous in other respects. That is just what our view of theoretical virtues implies.

4. A Brief Illustration

It may be helpful to illustrate the Tri-Level Method in action. We've already mentioned the contest between solipsism and the external-world view, in which the latter comes out on top: since solipsism fares so poorly at the first level, it fails to qualify as even minimally adequate, despite its simplicity. (Things only get worse at the second level.) What about less straightforward disputes, such as those concerning skepticism, free will, morality, or the mind–body problem? Given their complexity, and our efforts to keep this book short, any illustration involving such disputes must be schematic.[11]

Our illustration focuses on two prominent theories that endeavor to resolve the mind–body problem—that is, to answer the question of the connection between mental states and bodily states. According to the identity theory, each type of mental state is identical with a type of brain state; by contrast, dualism maintains that mental states cannot be wholly accounted for by physical states, including brain states.

How does each theory fare with respect to accommodating the data, the Tri-Level Method's first criterion at level one? Both the identity theory and dualism seem well-positioned to draw on claims that would accommodate data about conscious experience and attitudes such as belief and desire. Such data might include:

(a) attitudes such as belief and desire are intentional;

(b) many mental states bear only contingent connections to behavior;

(c) many mental states bear necessary connections to one another.

There are also candidate data that, it is often claimed, dualism is able to accommodate but the identity theory is not, such as:

(d) there is "something it is like" to have conscious experience;

(e) mental states are multiply realizable.[12]

---

[11] For those with an interest in metaethics and time on their hands, we offer two volumes in which we put the Tri-Level Method fully to work in the course of constructing and evaluating versions of non-naturalist moral realism and moral intuitionism. See Bengson, Cuneo, and Shafer-Landau (2023 and forthcoming).

[12] This datum has been said to be collected using thought experiments and the tools of empirical science (see, e.g., Putnam 1967 and Block and Fodor 1972).

On the other side, there are candidate data that the identity theory alone is said to be poised to accommodate, such as:

(f) in human beings, as well as in diverse non-human animals, all mental states are correlated with various kinds of brain activity;

(g) there are causal connections between mental events and physical events.

Identity theorists can respond to the charge that they are unable to accommodate (d) and (e) either by questioning these data or by supplementing their theory with weighty claims designed to offer the needed accommodation. For example, identity theorists could deny (e) by offering arguments designed to cast doubt on the phenomenon of multiple realizability.[13] Or they might supplement their core view with substantive claims about the nature of qualia, affirming a reductive account that makes (d) likely.[14] Dualists can respond in similar fashion. They might deny (f) and (g)—say, by endorsing epiphenomenalism or occasionalism[15]—or supplement their theory with ambitious claims designed to accommodate these data. Indeed, some dualists have suggested positing various psychophysical laws to accommodate (f), and others have offered dualist-friendly accounts of mental causation to accommodate (g).[16]

---

[13] See, e.g., Zangwill (1992) and Shapiro (2000).

[14] See, e.g., several of the contributions to Gozzano and Hill (2015).

[15] Jackson (1982) famously opts for epiphenomenalism, while Malebranche (1674/1997, 225, 448–50, and 660–70) endorses occasionalism.

[16] See, e.g., Yablo (1992) and Bealer (2007), who propose alternatives to Descartes' (now-defunct) claim that non-physical minds interact with physical bodies via the pineal gland, as depicted in Illustration 7, which appears in the 1664 publication of Descartes' *Treatise on Man*.

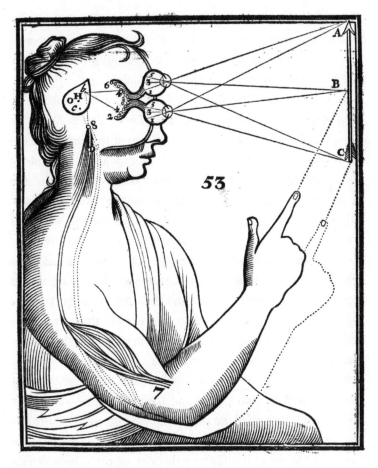

Illustration 7 Descartes on mind–body interaction.

Turning next to explanation of the data, called for by the Tri-Level Method's second criterion, both theories must be supplemented to explain a range of data, including (a)–(c). Some of the other data have been thought to pose more significant challenges. For instance, according to proponents of the "explanatory gap," the identity theory is in principle incapable of explaining (d); identity theorists demur, and may offer their own supplementations designed to provide the requisite explanation (or adequately defend the claim that no explanation is required).[17]

At level two of the Tri-Level Method, we find both theories in need of defense, the first component of substantiation. While we recognize that theorists have not endeavored to conform to the Tri-Level Method, the method incorporates familiar elements of good philosophical activity. Here, that amounts to the provision of arguments meant to constitute positive reasons for belief. Arguments implying that dualism alone is consistent with data concerning self-consciousness, certainty, and the possibility of zombies, disembodiment, inverted spectra, or dancing qualia are given as reasons to believe dualism, whereas inductive arguments from the past success of science and abductive arguments centering on the premise that identities best explain the correlations and connections in (f) and (g) are offered in support of the identity theory. As for the second component of substantiation, which concerns explanation, whether dualism and the identity theory can explain their own claims and commitments is sometimes questioned (though for the sake of brevity, we won't attempt to rehearse these challenges here). The debate over integration is fairly straightforward: while both theories appear to be internally coherent, dualism is alleged to conflict with our best science, whereas the identity theory is alleged to conflict with common sense. Accordingly, proponents of each theory will find themselves needing either to show that their theory is in fact compatible

---

[17] Levine (2001) presses the "explanatory gap."

with our best picture of the world, or instead to adequately defend the innocence of any such conflict.

Finally, moving to level three and its emphasis on theoretical virtues, there has been some debate over whether dualism or the identity theory is simpler; however, most participants in the debate appear to recognize that there is enough going on at the first two levels to render this particular issue moot, or at least far from decisive. This would make sense if theoretical virtues such as simplicity are mere tie-breakers, as our version of the Tri-Level Method maintains.

As this brief illustration indicates, conforming to the method would not represent a radical reorientation of how to engage in philosophical theorizing; the activities that it instructs theorists to engage in are familiar. At the same time, the illustration also helps show that conforming to the method would *not* be to engage in business as usual. Here we note a few of the distinctive features of the method.

First of all, while views such as the identity theory and dualism are radically different, the Tri-Level Method calls upon each theory to handle the full range of data, which are common currency between them. So the method rules out approaches according to which theorists cherry-pick data, accommodating and explaining data when their theory is well-suited to doing so, while downplaying or ignoring the data that remain. Instead, according to the method, when it comes to recalcitrant data, a theory must move in one of two directions. Either it must search for supplements that will accommodate and explain the data that it itself cannot, or it must adequately defend the claim that these data needn't be accommodated and explained, by (say) providing the relevant data disablers. Otherwise it must acknowledge that it's stuck with burdens that it fails to discharge—remainders that indicate how far the theory falls short of delivering theoretical understanding of its subject matter.

In addition, the Tri-Level Method instructs theorists to craft their views as well and fully as they can, rather than to simply exchange arguments and objections with their rivals, as sometimes happens when theorists merely shift burdens of proof or fixate on admissible moves within a given dialectic. While arguments and replies to objections have a place (as part of defense), the method applauds theories that accommodate and explain the data, and substantiate and integrate their own claims and commitments (or, alternatively, adequately defend the claim that they needn't do one or more of these things). Importantly, doing all this is not just a comparative venture, in the sense that it does not involve theorists—whether proponents of dualism, the identity theory, or some other view—supporting their claims, commitments, and argumentative strategies merely by establishing that they compare favorably to those of rival theories. Nor does it consist in theorists looking over their shoulder at those rivals, canvassing and responding to objections they might pose. Nor is it a matter of inquirers establishing that their view exhibits theoretical virtues that competing positions do not. In this way, the method instructs philosophers to focus not on other theories, but rather on handling the data, first and foremost.[18] The view that satisfies the method's criteria to a high degree will be poised to deliver

[18] Castañeda (1980, 133) similarly urges philosophers to focus on the data, as opposed to (say) engaging in argumentative hot potato or making interventions in local disputes.

understanding of its subject matter, whatever else other theories might or might not accomplish.

Finally, the Tri-Level Method's criteria are non-partisan in (at least) the following two respects. First, they do not prefer the theory that fits with some comprehensive philosophical view, such as naturalism, theism, panpsychism, or the like. For such views do not possess default status, and they do not belong to our best picture of the world, in our semi-technical sense of that notion. Second, the method's criteria do not privilege views that exhibit one or another theoretical virtue, such as parsimony or unification. Theoretical virtuosity is relevant only when all else is equal. A view such as the identity theory is free to emphasize its simplicity. Likewise, a view such as dualism may emphasize that its central notions (e.g., pain) are joint-carving. But both theories must look elsewhere to earn the claim to enhancing our understanding of the relation between mind and body. In short, they must do this by accommodating and explaining the data, and by substantiating and integrating their own claims and commitments (or, alternatively, by satisfying the corresponding escape clauses).

5. Answering the Question of Method

We defined a method as a set of criteria for theory construction and evaluation. The directions implied by the Tri-Level Method's constituent criteria supply substantial practical guidance for theory construction. These criteria say what theorists need to do in order to develop a theory that handles the data, is well-grounded, and (as relevant) is theoretically virtuous. The method also provides a means by which to determine a theory's merits: it is assessable in light of how well or poorly it fares with respect to the method's five criteria. This is crucial for theory evaluation.

Perhaps more importantly, the Tri-Level Method handles all three data about method collected in the previous chapter. First, the method accommodates and explains the friendliness datum regarding central features of theoretical debate. For the method is friendly to the activities in which philosophers engage when they're doing philosophy, such as

- advancing arguments
- raising objections
- offering replies to these objections
- providing clarification
- developing explanations
- displaying sensitivity to the deliverances of logic, mathematics, science, and (to a greater or lesser degree) common sense.

The Tri-Level Method also explains why it is a *good* thing for philosophers to engage in those activities: doing so conduces to the satisfaction of the method's criteria, which are constitutively linked to inquiry's ultimate proper goal. By advancing arguments and raising or replying to objections, philosophers improve their theories through stress tests that produce primary and secondary defenses.[19] Developing explanations helps to satisfy the other

---

[19] The Tri-Level Method also accommodates and explains more specific aspects of these elements of philosophical practice, such as the nature of objections (see the Box below) and the importance of and limits to arguments (see, e.g., Nozick 1981, 4; van Inwagen 2006, Lecture 3; and Ballantyne 2013) and counterexamples (see, e.g., Weatherson 2003 and Bonevac, Dever, and Sosa 2011). For instance, counterexamples are important because they expose failure of accommodation; they are limited because the Accommodation Criterion can be satisfied even if some of the data are not accommodated—as permitted by its escape clause.

portion of the Substantiation Criterion, or instead the Explanation Criterion. Sensitivity to the deliverances of logic, mathematics, science, and common sense positions theorists to fulfill the Integration Criterion. As for clarification, this could occur at any point. In some cases, it might be as simple as elucidating key terms in claims advanced at levels one or two, or being explicit about the implications of those claims. In other cases, the needed clarification may be more complicated, as when theorists sharpen a view by distinguishing it from close competitors, or by anticipating and forestalling misinterpretations. Clarification might even come in the form of an analysis, which may in some cases accommodate and explain a range of data, thereby serving as the basis of a powerful theory that answers a variety of questions, including those of the form "What is F?" and "Why is F distinct from G?."[20]

The nature and variety of objections

The Tri-Level Method suggests a simple and informative characterization of the notion of an objection:

> A consideration is an *objection* to a philosophical theory if, only if, and because that consideration provides reason to think that the theory does poorly in some regard with respect to one or more of the Tri-Level Method's criteria.

The method also sheds light on familiar styles of objections and why they qualify as objections. For it identifies which dimension of theory evaluation each could be reasonably taken to target. To illustrate:

---

[20] We hope these remarks go some way toward indicating that a lot of contemporary philosophical work contributes to the satisfaction of the method's criteria, whether directly or indirectly. Insofar as it does, such work is of considerable value.

- *Not really a theory, incomplete*: The theory does not accommodate and explain sufficiently many data (or adequately defend this lacuna), and so does poorly with respect to the Accommodation and Explanation Criteria.

- *Descriptively inadequate, counterintuitive, phenomenologically off-key*: The theory does not accommodate sufficiently many data (or adequately defend this lacuna), and so does poorly with respect to the Accommodation Criterion.

- *Explanatorily inadequate, impoverished, unilluminating*: The theory does not explain sufficiently many data (or adequately defend this lacuna), and so does poorly with respect to the Explanation Criterion.

- *Ad hoc, unmotivated*: The theory does not defend one or more of its own claims or commitments (or adequately defend this lacuna), and so fares poorly with respect to the Substantiation Criterion.

- *Mysterious, stops too soon*: The theory does not explain one or more of its own claims or commitments (or adequately defend this lacuna), and so fares poorly with respect to the Substantiation Criterion.

- *Circular*: The theory explains one or more of its own claims or commitments by invoking one or more data that those claims or commitments are intended to explain, and so does poorly with respect to the Substantiation Criterion.

- *Untenable, implausible, contravenes science or common sense*: The theory conflicts with our best picture of the world (without adequately defending the claim that this conflict is unproblematic), and so does poorly with respect to the Integration Criterion.

- *Unparsimonious, inelegant, ugly, unruly, unnatural, intractable, etc.*: The theory is in some non-trivial respect less theoretically virtuous than rivals, and so would do poorly with respect to the Virtue Criterion, were it applicable.

We submit that these are illuminating characterizations of precisely the sorts of objections that philosophers legitimately attempt to head off when constructing their theories, or advance against rival theories in their chosen domains.

Second, the Tri-Level Method's five criteria are determinative: they select among multiple theories that fit the data, eliminating a great many from contention. The method thus resolves the underdetermination of theory by data. For it identifies a set of instructions that, when successfully followed, allow inquirers to zero in on a theory that does not merely imply or probabilify the data, but provides understanding to anyone who fully grasps it.

Third, the Tri-Level Method handles the virtues datum, exhibiting the features of a sound method identified in Chapter 4 (§2). The method is *comprehensive*: its Accommodation and Explanation Criteria call for theories to address not just a single type of consideration, but a wide range of data within the target domain. It is also *support-requiring*: it grants no theory "presumptive" or "default" status, since a theory earns its positive status only if is well supported by reasons, per the defense component of the Substantiation Criterion. In addition, the method is *synoptic*: it calls for explanations of both the data and the theory's own claims and commitments, per the explanatory component of that criterion. It thereby ensures that whichever theory it favors is best positioned to expose systematic relationships in the subject matter in question. Moreover, the method is straightforwardly *multidimensional*: it does not privilege a particular theoretical desideratum, but rather identifies at least four (accommodation, explanation, substantiation, and integration) and, as relevant, a fifth (virtuosity) that inquirers must attend to in theory construction and evaluation. Furthermore, the method is *hierarchical*: it assigns these desiderata diverse roles, thereby acknowledging that some desiderata possess more significance than others (in particular, the Virtue Criterion is in most cases going to be far less important than the criteria at levels one and two). Finally, the Tri-Level Method is *comparative*: it enables assessment of the individual as well as the relative merits of rival theories. Each candidate can be evaluated in light of the extent to which it satisfies the method's criteria, and competing theories can be measured against one another by reference to how well they do this.[21]

Our contention, then, is that the Tri-Level Method handles all three of the data regarding method. When setting forth its criteria, we pointed out how they are constitutively linked to the six features of a theory that provides understanding. Accordingly, we submit that the method is sound, yielding a satisfactory answer to the question of method. Earlier we observed that if a method is sound, then (ceteris paribus) inquirers have strong reason to endorse its outputs. This is why—despite the high bar that it sets—we encourage adoption of the Tri-Level Method in pursuit of answers to philosophical questions, and of philosophical progress more generally.

---

[21] Recall that all of the extant methods discussed in Chapter 4 (§3) lack at least one of these virtues. We've also argued that those methods fail to position inquirers to secure outputs possessed of the six features of theoretical understanding, and many of those methods are not obviously determinative or friendly to all of the commonplace philosophical activities discussed in Chapter 4 (§1). These are further reasons for relying on the Tri-Level Method, rather than any of these familiar alternatives, when seeking to make philosophical progress (our topic in the next chapter). Independent support may be found in Chalmers' (2014, §§4–5) and Beebee's (2018, §II) suggestions that the alternatives (in particular, the Method of Argument and Cost-Benefit Method, respectively) impede rather than facilitate progress.

6

Philosophical Progress

There is a longstanding and influential charge that philosophy's progress does not measure up to that made in mathematics and the natural sciences. Bertrand Russell famously pressed the worry in *The Problems of Philosophy*:

> If you ask a mathematician, a mineralogist, a historian, or any other man of learning, what definite body of truths has been ascertained by his science, his answer will last as long as you are willing to listen. But if you put the same question to a philosopher, he will, if he is candid, have to confess that his study has not achieved positive results such as have been achieved by other sciences.[1]

Here Russell characterizes the "positive results" of the sciences and philosophy as a "definite body of truths," rather than as the

---

[1] Russell (1912/1997, 154–5). Notice that this passage does not distinguish between truths "ascertained by his science" that are *data* and those that are part of a *theory*. To ensure a level playing field, we will focus on theories in what follows.

Philosophical Methodology: From Data to Theory. John Bengson, Terence Cuneo, and Russ Shafer-Landau, Oxford University Press. © John Bengson, Terence Cuneo, and Russ Shafer-Landau 2022. DOI: 10.1093/oso/9780192862464.003.0007

achievement of understanding, inquiry's ultimate proper goal. But this characterization is incidental to his central negative thesis, namely, that philosophy compares unfavorably with other intellectual fields in respect to progress. This thesis continues to be widely accepted.

Russell's doubts are emblematic of a more basic worry about philosophical progress, which focuses not on the alleged inferiority of philosophy vis-à-vis "the other sciences," but instead on the ability of philosophy to register any progress at all. Kant's anxieties on this score are well-known. Speaking of what he calls "metaphysics," Kant writes that it is

a battlefield, and indeed one that appears to be especially determined for testing one's powers in mock combat; on this battlefield no combatant has ever gained the least bit of ground, nor has any been able to base any lasting possession on his victory. Hence there is no doubt that up to now the procedure of metaphysics has been a mere groping....[2]

Kant's successors have extended his pessimism well beyond the boundaries of metaphysics to philosophy as such.

Why has progress proven so elusive? Kant intimates that it is due to the failure to develop and implement an adequate methodology. He is not alone; many have pointed to method as the culprit. The idea here is not simply that philosophers have failed to articulate, or properly adhere to, their methodological commitments: this by itself is not the impediment to progress that troubled Kant and others. Nor is the worry that philosophical debates are rife with disagreement—in our view, consensus is not always a sign

---

[2] Kant (1787/1998, Bxv 109–10).

of a discipline's health, whereas meticulous debate often is.[3] Rather, the deep problem concerns the absence of any method whose proper execution is likely to further understanding regarding philosophy's central questions. Such progress, which we'll call 'noetic,' would thus seem beyond reach.

Focusing on noetic progress is appropriate not only because theoretical understanding is inquiry's ultimate proper goal. Such a focus also has the virtue of recognizing that philosophical investigation needn't be assessed in terms of its success at problem-solving. In fact, even if one or more of the leading questions about a given domain is unsolvable, this would not by itself prohibit understanding regarding that domain. Noetic progress is more complicated, and more nuanced, than that.

These remarks are intended to highlight the various layers of the problem of philosophical progress, an adequate solution to which must address four questions:

Existential Question: Has philosophy actually made noetic progress with respect to its central questions?

Comparative Question: Is there noetic progress in philosophy with respect to its central questions that is comparable to progress made by other fields of theoretical inquiry?

---

[3] *Cp.* Solomon (2001), Beatty and Moore (2010), and Dang (2019). It may be that extensive disagreement poses an additional threat *if* it generates a defeater in the way described by conciliationist views (see, e.g., Elga 2007 and Feldman 2007; for applications to philosophical progress, see, e.g., Kornblith 2013, Goldberg 2013, Frances 2015, Lycan 2019, and Beebee 2018, §III). We are not convinced that philosophical controversy generates such a defeater. Together with the point we make in the text, this leads us to focus not on disagreement but on the alleged paucity of philosophical progress itself. We return to the topic of disagreement in §2.2.

How-To Question: How is it possible for philosophy to make noetic progress with respect to its central questions?

Evaluative Question: How is it possible to attain a principled evaluation of philosophy's noetic progress (if any) with respect to its central questions?

These questions are related in interesting ways. Answering the second pair requires the provision of a method that can guide the construction of philosophical theories poised to furnish theoretical understanding, and to enable a principled assessment of the extent to which they deliver this good. If our method-oriented diagnosis above is correct, providing such a method is also required to answer the first pair of questions in a fully satisfying way. So, although the existential and comparative questions have received the bulk of the attention in recent literature, we submit that the how-to and evaluative questions identify the root of the problem.[4]

In this chapter, we'll discuss how the Tri-Level Method positions us to offer answers to all four questions, beginning with the second pair.

1. The How-To and Evaluative Questions

To appreciate how the method assists with addressing the how-to and evaluative questions, it will be helpful to distinguish three interlocking but distinct intellectual projects.

Theory construction consists in developing a candidate theory of a given domain by articulating a set of claims about the data in that domain that satisfy the Accommodation, Explanation,

---

[4] Many discussions of philosophical progress omit one or more of the questions we've identified. Chalmers (2014, 3) helpfully distinguishes the first pair while also diagnosing the root problem as one of method.

Substantiation, and Integration Criteria (and the Virtue Criterion, as relevant). *Theory appraisal* consists in applying these criteria to a theory in order to determine how well it fares with respect to each of the method's first two levels. *Theory comparison* consists in placing appraisals of multiple theories side by side to determine which view fares best with respect to the criteria at the first two levels (and the Virtue Criterion, in case of a tie).

The Tri-Level Method makes it possible for inquirers to achieve noetic progress on each of these intellectual projects. Our answers to the how-to and evaluative questions explain how.

1.1   Answering the how-to question

The how-to question focuses on theory construction, asking after philosophy's ability to develop theories that make noetic progress with respect to its central questions. Our answer should by now be clear: by constructing a theory of a domain that satisfies the four criteria at levels one and two to a high degree, and (as relevant) the fifth criterion at level three.

We hasten to emphasize that when satisfying these criteria, inquirers may at various points need to draw upon work done by theorists working in other areas. This sort of division of labor, in which a theorist relies on the contributions of others to help execute the method's instructions, is not illicit, so long as theorists are forthcoming about the precise junctures at which they do this. Because so much of philosophy is interconnected, because the call for explanation and defense can extend almost indefinitely, and because a showing of internal and external integration can be such a complicated matter, it is perfectly legitimate for theorists to cite the fruits of others' labor in lieu of providing complete substantiations and integrations themselves. It is a virtue of our preferred method that inquirers are encouraged to lean on each other in this way (and to acknowledge when they do so).

Our answer to the how-to question leaves much open. In particular, the Tri-Level Method does not itself drill down and include precise instructions on how best to engage in theory construction. Theorists are called on to accommodate and explain the data; they are instructed to substantiate and integrate their claims; they must, as needed, craft views with an eye to their virtuosity. But the method itself does not dictate specific ways in which to do these things. (For example, as noted in Chapter 5 (§2.3), it may be useful to take up the levels in order. But inquirers may find cause to revert to the first at any time; after all, handling the data is normally an ongoing venture.) Being somewhat open here is an advantage, as it offers theorists flexibility in their constructive efforts, allows for a diversity of legitimate paths to theory construction, and spurs the creativity of inquirers to assemble theories that are sensitive to the method's constituent criteria. While our presentation of these criteria is compatible with many different ways of filling in various details, we can afford to remain neutral on them, as they don't affect the basic shape of our answer to the how-to question (or indeed any other aspect of the problem of progress).

1.2　Answering the evaluative question

While the how-to question is directed to matters of theory construction, the evaluative question focuses on theory appraisal and comparison, which are both facets of evaluation. How is it possible to attain a principled evaluation of philosophy's noetic progress (if any) with respect to its central questions? By once more following the Tri-Level Method: applying its criteria in order to determine the extent to which a theory of a given domain has accommodated and explained the data in that domain, substantiated and integrated its claims and commitments, and (as relevant) realized various theoretical virtues. We'll first discuss how the method helps with theory appraisal before turning to its treatment of comparative evaluation.

When setting out the method's criteria in the previous chapter, we proposed that a theory is minimally adequate just in case it does modestly well when it comes to the criteria at level one, and respectable just in case it also does modestly well regarding the criteria at level two. Other categories of appraisal are possible. For example, an ideal theory, which satisfies all of the Tri-Level Method's criteria to a maximal degree, would bestow an extraordinary sort of understanding. But a theory need not be ideal to be excellent, in the sense of supplying understanding of the target domain that is both substantial and such that further constructive efforts are likely to foster only modest improvements in that understanding. For us imperfect inquirers, an excellent theory is an eminently worthy goal.

We recognize that philosophers who agree about the data sometimes disagree about what claims yield accommodation and explanation. They also sometimes disagree about what an adequate defense or explanation would be, or which claims call for explanation. Whether one claim coheres with another is sometimes contested; whether it fits with our best picture of the world, and what the elements of that picture include, are subject to yet further dispute. Such controversies are natural. But they don't threaten either the letter or the spirit of the criteria we've identified so far. In fact, there is a sense in which these criteria are often presupposed in such controversies.

Granted, these controversies can pose a practical challenge for the project of appraisal. But the difficulties are mitigated to the extent that there is widespread agreement on a broad range of issues. To illustrate, epistemologists have for the most part shared the view that any theory of a priori knowledge should address at least the following topics:

(1) how we come to possess such knowledge;
(2) whether it is empirically defeasible;
(3) whether it can be synthetic and not merely analytic; and
(4) the extent to which it coheres with our best picture of the world.

To be sure, there is sometimes disagreement about which of these items deserves the first look when developing a view of the a priori. Many skeptics of the a priori emphasize (2) and (4), while its proponents tend to focus on (1) and (3).[5] Though these disagreements are important, they should not obscure the fact that there is very considerable agreement that a satisfactory theory must somehow address all four.

Furthermore, it is now widely accepted that some influential attempts to say what a theory's claims must be like miss the mark. Take, for example, Paul Benacerraf's suggestion that any view that commits itself to a priori knowledge of mathematical truths must explain this commitment by identifying a causal relation between

---

[5] Skeptics include Quine (1951), Kitcher (2000), and Devitt (2014); proponents include Bealer (1992), BonJour (1997 and 2014), Chalmers (2002), and Peacocke (2004, Ch. 6).

the knower's beliefs and the mathematical facts they target.[6] For a variety of reasons, few regard this requirement as satisfactory.[7] The presence of such reasons should help to assuage concerns about the feasibility of applying the method's criteria.

Still, we acknowledge that appraising a theory is no mean feat, as it will require assessing how well or poorly the view has done by the lights of all four criteria at levels one and two. The Tri-Level Method is officially neutral on how best to render such appraisals. For example, it takes no stand on how best to measure the adequacy of explanations; it provides no algorithm for determining that one defense is superior or inferior to another; it does not itself say how to gauge coherence. While various meta-methods could be introduced to address these further questions, doing so would not affect the constitution of the Tri-Level Method itself, but would instead represent elaborations and refinements of it. Even in the absence of such meta-methods, however, the four criteria and their ordering are valuable tools for theory appraisal.

As noted, theory comparison follows upon such appraisal. For theorists can put multiple views shoulder to shoulder only after first assessing each individually. Whichever theory satisfies the four criteria at levels one and two to the highest degree is best; such a theory is on track to provide the greatest understanding of its subject matter. In the unlikely event that multiple theories fare roughly equally well at those levels, the Virtue Criterion will select

[6] Benacerraf (1973, §IV).

[7] One reason is that it seems to conflict with inductive knowledge and knowledge of the future; another is that Benacerraf and others insufficiently motivate the thesis that the relevant explanation must be causal. The discussion of the matter is substantial. See, among others, Burgess and Rosen (1997, 46–9), Linnebo (2006), and Bengson (2015).

the one that comes out on top.[8] The upshot is a principle of inference that we'll call

> **Inference to the Understanding-Provider (IUP):** If a respectable theory satisfies the Tri-Level Method's criteria to the highest degree relative to its rivals, then (ceteris paribus) inquirers have strong reason to accept that theory over those rivals.

This principle is a natural consequence of our contention that the Tri-Level Method is sound (i.e., is poised to deliver understanding), together with our earlier observation that inquirers have strong reason to endorse the outputs of a sound method.[9]

We acknowledge the difficulties attaching to the sort of comparison required to apply the IUP. Theories will differ as to how they accommodate and explain the data, or elect to satisfy the call for substantiation and integration. So they will also differ as to their claims and commitments. Such differences may be numerous, and will in many cases be quite radical. Once we recognize these points, the project of comparative evaluation can seem overwhelming.

But a counsel of despair is probably premature. While it is true that determining the comparative merits of a given theory will not always be straightforward, this is not to say that it is always impossible. On the contrary, there are often clear cases in which we are

---

[8] What happens in the (highly unlikely) case that there is *also* a tie with respect to the Virtue Criterion? In such a scenario, we must rest content with a tie when it comes to the question of which theory best promotes understanding. However, we're open to the possibility that in such a context other features of theories (e.g., their practical benefits) might be invoked to settle the different question of which view to accept.

[9] One neat feature of the IUP is that it promises to justify and explain a version of the principle known as inference to the best explanation (IBE): if a theory is better than its rivals at explaining some phenomena, then (ceteris paribus) inquirers have strong reason to accept it over those rivals.

able to make such determinations. Consider a minimally adequate theory that defends, explains, and integrates all the claims and commitments it makes at the first level. It is clearly likelier to promote understanding than another theory that fails to handle the data, or does so while leaving its claims and commitments unsubstantiated and unintegrated (without utilizing the escape clauses). The first theory is respectable; the second is anything but.

Successfully rendering a comparative evaluation will depend on numerous factors, several of which we highlighted when noting various meta-methods that could prove handy for appraisal. Theorists might wish to reach for additional meta-methods to help them fulfill the task of theory comparison. While the Tri-Level Method does not oppose doing these things, it also does not require them. In any case, they are inessential to our project here. What we are doing is presenting a method that locates the sorts of criteria whose satisfaction by a theory implies that it is best poised to advance understanding of a subject matter. We can say, in general terms, what the key notions of accommodation, explanation, substantiation, integration, and virtuosity amount to, as well as how they interact. It is no part of our brief to defend a single, and doubtless contentious, specification of those notions, of the kind required to formulate and implement the relevant meta-methods. Moreover, recall that we have argued that the Tri-Level Method's criteria are constitutively linked to theoretical understanding. Importantly, this link will remain on any plausible elaboration and refinement of the method's key notions, and will be preserved by any legitimate meta-methods. This is one reason we find it unnecessary to dwell on these matters here.[10]

---

[10] Even as presented, the Tri-Level Method provides an informative and useful guide. In point of fact, we've already witnessed two substantive applications of its criteria:

2. The Existential and Comparative Questions

We've argued that the Tri-Level Method helps to reveal how noetic progress is possible and how philosophical theories can be evaluated with respect to their ability to facilitate such progress. The method thus plays a crucial role in answering the how-to and evaluative questions. Our answers to these questions were designed to show that, and explain how, noetic progress is possible. But they do not by themselves resolve the existential and comparative questions. What remains is to address whether philosophy has actually made any noetic progress, and if so, how it compares to the leaps and bounds made by the natural sciences. We also need to ask whether the Tri-Level Method helps to explain any such progress.

2.1 Fundamental vs. foremost questions

We believe pessimism about philosophy's noetic progress to be unwarranted, being based on an illicit slide between two importantly different types of questions. The first are the *fundamental* questions of a field, whose resolutions are among the field's ultimate (perhaps even constitutive) aims at any time. The second are a field's *foremost* questions at a specific time, which are not. The

---

First, we canvassed data about data (recall the four basic features of data in §1 of Chapter 2), and subsequently showed that the epistemic reasons theory accommodates and explains all of them. Second, the method also applies to the choice of method (hence, to theorizing about philosophical method itself); this is indicated by our efforts to collect data about method in Chapter 4 (§§1–2), and our discussion of their handling by the Tri-Level Method in Chapter 5 (§5; see also its note 21). As these two cases demonstrate, the Tri-Level Method is not only efficacious but also has broad scope, applying wherever there are data that constrain philosophical inquiry.

distinction is intuitive and can be illustrated by calling to mind questions from non-philosophical disciplines.

Consider the natural sciences: On the one hand are questions concerning, say, the origins of the universe, the basic building blocks of matter, and the earliest genesis of life. On the other hand are questions concerning such things as the existence and character of black holes, the consistency of principles of special relativity and quantum mechanics, and the character of the human genome. Although those on the latter list are of great importance, and can be counted among the foremost questions in contemporary scientific investigation, they are not fundamental to their respective fields in the same way as those on the former list. The distinction also has application in mathematics, where the fundamental questions arguably concern, *inter alia*, the structure of the natural numbers and the distribution of primes. By contrast, the twenty-three mathematical problems famously articulated by David Hilbert in 1900 delineate not the fundamental questions of mathematics, but rather (in Hilbert's own words) specific questions "which the science of to-day sets and whose solution" is—or, we should now say, was—of foremost interest and importance, bearing "deep significance…for the advance of mathematical science" in the twentieth century.[11]

Philosophy, too, has its fundamental questions, some of which were retailed at the beginning of this book. Each of these questions belongs to and helps to shape the contours of particular subfields within philosophy.[12] While these questions are certainly on

---

[11] Hilbert (1902, 437–8).

[12] Fundamental questions do not, however, determine the domains or subject matters that those subfields investigate. So such questions differ from what Stoljar (2017, 12) calls "topic questions," which "introduce or define or constitute a topic or subject

the radar of contemporary philosophers, much philosophical research in each of these subfields over the past few decades is best understood as devoted to far more specific questions, which are not fundamental in the same way, but are plausibly classified as the foremost questions at the time. These include questions about the relation between consciousness and behavior (philosophy of mind), the varieties of knowledge (epistemology), the basic features of the grounding relation (metaphysics), the distinction between constraints and permissions (normative ethics), the relation between semantics and pragmatics (philosophy of language), and the connection between moral reasons and motivation (metaethics).

It should be clear that when it comes to the fundamental questions of the natural sciences, mathematics, and philosophy, inquirers have not yet achieved the level of theoretical understanding to which they aspire. At the same time, we maintain, inquirers in all these fields have achieved substantial understanding with respect to many of the foremost questions they tackle. Though the progress in philosophy may not be equivalent to that made in the natural sciences and mathematics, once we compare apples to apples, and look at the respective progress made on foremost questions, we can see that philosophy is not the great disappointment that Russell made it out to be.[13]

---

matter." Rather, fundamental questions are among the biggest of what Stoljar (2017, 11) calls "big questions," though most foremost questions are also big in Stoljar's sense.

[13] At least one recent defense of Russell's negative thesis explicitly contrasts mathematicians' successes (or partial successes) regarding the majority of Hilbert's questions with lack of success regarding the philosophical questions treated in Russell's text (see Chalmers 2014, 9–12). But as we've just explained, Hilbert's questions are foremost, not fundamental. All, or nearly all, of the questions in Russell's text—regarding the distinction between appearance and reality, the possibility of knowledge, induction and the a priori, universals, and the nature of truth—are fundamental. Other, similarly illicit

The distinction between fundamental and foremost questions not only speaks to the comparative question; it also helps to address the existential one. For it points to a flaw in a natural line of reasoning that would appear to lead inexorably to the conclusion that philosophy has made no progress. The reasoning has three steps. The first maintains that a field makes genuine progress just to the extent that, over time, it supplies answers to its central questions. The second observes that the central questions of philosophy are among life's "big questions," concerning such topics as God, free will, the mind–body relation, knowledge, and morality. Step three delivers the bad news: we lack answers to *any* of these questions.

In our view, this reasoning should not convince—and not simply because inquiry's ultimate proper goal is understanding (rather than the mere provision of answers). We ought to reject step one if 'central questions' is restricted to fundamental questions, and step three if 'central questions' is restricted to foremost ones. Either way, the anti-progress conclusion in this reasoning is averted.

That reply is successful, however, only if philosophy has indeed achieved considerable understanding with respect to its foremost questions. After defending this claim, we'll bring in the Tri-Level Method to help explain such progress. The overall result will be an answer to the existential question that provides the raw materials for resolving the comparative question.[14]

---

comparisons are made by, e.g., McGinn (1993, 12), Dietrich (2010, §2), Gutting (2016, §11), and Papineau (2017).

[14] The modest position we'll develop contrasts with pessimistic views of progress that emphasize our inherent cognitive limitations or aporetic features of philosophical questions (*cp.* McGinn 1993 on "cognitive closure" and Nagel 1986 on "points of view"). Our position, though hardly pessimistic, also contrasts with Stoljar's (2017) highly

2.2 Shared frameworks

Recall that convergence on a theory is not a precondition of its ability to yield understanding; some inquirers may well fail to notice a theory's noetic promise. What is more, philosophy is so difficult that it is rational for inquirers to pursue divergent views. Spreading out in this way allows them to cover more terrain.

Still, we recognize that some may be inclined to see persistent philosophical disagreement as defeating claims to noetic progress. We encourage those with such an inclination to observe that there is substantial convergence in philosophy, whose extent frequently goes unnoticed. Philosophers thinking self-consciously about their discipline are often attuned to the debates therein. But they shouldn't overlook the vast agreement that subserves those debates. This agreement operates as a common framework, whose elements constitute a partial view of the domain under investigation. The (typically controversial) theories at the center of philosophical debates build on these shared frameworks.

For instance, when philosophers disagree about the existence of God, or the nature of causation, or the principles of justice, they typically do so against the background of extensive agreement—"the axis around which the body rotates."[15] The framework that unites

---

optimistic view that philosophers have resolved many "big questions" (see note 12) through answers that simply *replace* the original question with a successor. Since in each case the original and successor are arguably both instances of the inquiry-opening question (which is rightly labeled 'big'), we find this approach to philosophical progress unsatisfying. Our answers to the existential and comparative questions do not appeal to successors. In addition, while Stoljar (2017, 32) is silent on the issue of methodology, our approach to progress is accompanied by answers to the two questions about method—the how-to and evaluative questions—that identify the root of the problem.

[15] To coopt a phrase from Wittgenstein (1969, §152).

theorists in each area is bound to include a host of distinctions (e.g., between power and its exercise, objects and events, or actions and traits); an inventory of similarities and differences between the items that are distinguished; a set of necessary conditions for the target (e.g., God, causation, or justice), as well as various sufficient ones; information about these conditions themselves, such as how they can (and can't) be legitimately precisified; various claims regarding the target's explanatory role; a record of theoretical dead-ends; a roster of open possibilities; and so on. Such agreements often slip through the cracks, with disagreements receiving the lion's share of attention.

Things appear to be reversed when it comes to the natural sciences and mathematics: there, a shared framework is what is most salient, and disputes are taken to be the reserve of specialists. This surmise is motivated by the observation that when people query progress in philosophy, they tend to emphasize the absence of answers regarding its fundamental questions. This tendency misses the possibility that there have been significant advances with respect to foremost questions; such advances, we submit, enrich the frameworks that form the backdrop against which robust philosophical controversies are sustained. By contrast, when people extol progress in the sciences, they often refer to the many advances regarding its foremost questions, passing over the fact that the fundamental ones remain open.[16]

This familiar dynamic rests on a double standard that is so pervasive as to be nearly invisible. If our shared frameworks hypothesis is correct, we can credit this double standard with responsibility

---

[16] There is of course considerable disagreement in the sciences regarding many foremost questions, too. See, e.g., Wolchover (2020) on physicists' views of particles.

for a sizeable portion of the pessimism about philosophy, and for a good deal of the invidious comparisons between its progress and that of other fields.

2.3 Forms of noetic progress

The foregoing remarks do not constitute answers to the existential and comparative questions, but do help to set the stage. The next step is to proffer compelling examples of noetic progress, which can take several forms.

First, understanding can be improved by identifying deep difficulties with claims or arguments that once seemed viable. Edmund Gettier's counter-examples deflated the JTB analysis of knowledge; logical positivism was undone by work that revealed the self-undermining character of its verifiability criterion; hedonism has been dogged by the specter of Robert Nozick's experience machine; endorsement of the logical problem of evil has largely disappeared, owing to Alvin Plantinga's exposure of its vulnerabilities; after the incisive analyses of Helen Longino and others, the longstanding notion that science is wholly value-free has breathed its last.[17] These are instances of manifest progress, even if there are pockets of resistance to the lessons to be taken from such work. For these negative conclusions sponsor positive views that, while continuing to be reason-based, robust, illuminating, orderly, and coherent, are now more accurate than the theories that were usurped.

---

[17] Because we take the examples in this paragraph and the rest of this subsection to be familiar to many readers, and in order to avoid overwhelming you with footnotes, we'll pass on our usual custom of offering citations.

Second, noetic progress can be made by articulating distinctions whose previous oversight hampered theorizing about a given domain.[18] Our understanding of the mind has substantially improved as a result of regimenting the difference between phenomenal and access consciousness. Prizing apart varieties of simplicity (e.g., ontological, qualitative, ideological, syntactic, etc.) promotes a better understanding of its role(s) in theory choice. The fourfold Hohfeldian division among types of rights and their correlative duties has assisted inquirers in cutting through various ambiguities and confusions when theorizing about these normative relations. Conceptions of free will have been enhanced by incorporating a distinction between agent- and event-causation. In each of these areas, theories attentive to such distinctions have a far greater chance of possessing one or more of the six understanding-enabling features.

Third, noetic progress can take the form of incrementally improving the viability of existing theories. There are countless examples to choose from. The proto-utilitarianism of Hume, or even the codified formulation offered by Bentham, is far less sophisticated and comprehensive than the consequentialist theory available in Sidgwick's *Methods of Ethics*. Similarly, today's versions of metaethical expressivism are light-years ahead of the non-cognitivist views they descend from. The philosophy of race has also developed in ways that render contemporary theories exponentially better than those contemplated in previous generations. Several factors make such innovations possible, including expansion of the "explanatory store" that philosophers are able to

---

[18] *Cp.* Gutting (2009 and 2016).

draw upon when refining their theories.[19] Recent theorizing has illuminated the contents of this store: relations such as supervenience, constitution, realization, composition, identity, grounding, and more can now be deployed in philosophical accounts. With these relations in hand, theorists can amplify and refine their views in ways that enhance understanding.

Fourth, noetic progress may occur by expanding the field of important possibilities in a given domain.[20] Examples here are also abundant, and include W. D. Ross' theory of prima facie duties in normative ethics, Ernest Sosa's and Linda Zagzebski's virtue epistemologies, Ruth Barcan Marcus' contributions to modal logic, and Martha Nussbaum's capabilities approach to well-being. We do not wish to pronounce on the veracity of any of these options. But it is safe to say that theories poised to respect the insights embodied by (or learn from the missteps in) such innovative approaches are more likely to be accurate, reason-based, robust, illuminating, orderly, or coherent.

All four forms of progress have broadened and deepened the sort of shared framework that is normally taken for granted in the course of philosophical theorizing. This framework is the backbone of an understanding-providing theory, albeit a partial one, of a given domain.

2.4 Illustrating noetic progress

Having identified at least four forms of noetic progress, let us now offer detailed illustrations of the possibility of noetic progress on

---

[19] We borrow the quoted term from Kitcher (1981, 512).

[20] *Cp.* Nozick (1981, 21–2) and Wilson (2014, 145–50 and 2017, 92).

some of the foremost questions in philosophy. Space limits us to two—one offered here, on the aforementioned connection between moral reasons and motivation, and the other, involving the philosophy of perception, in the Box that appears below. Each example centers on a coherent set of claims that represent the fruits of past theorizing. There is nothing special about the examples we've chosen; it would have been easy to identify many more.

Through a long process of critiquing once-prominent claims or arguments, making novel distinctions, incrementally improving leading contenders, and exploring brand new possibilities, contemporary metaethicists have unearthed the following non-obvious truths about the connection between moral reasons and motivation:

- Something is a motivation only if it is capable of contributing to an explanation of an agent's action.
- There is a distinction between (at least) two types of motivation: a certain type of mental state, such as desire, that plays a causal role in action (a 'motivator'), and a consideration in the light of which one acts (a 'motivating reason').
- Motivators and motivating reasons are both distinct from normative reasons, which are considerations that favor some response.
- A consideration is a good motivating reason if and because it is also a strong, undefeated normative reason.
- Something is a moral reason only if it is a normative reason.
- A reason is normative only if, when it is sufficiently strong, flouting it renders one blameworthy absent excuse.
- There is no direct correlation between the strength of a motivator and the strength of a normative reason.

- There can be a moral reason for an agent to act at a given time even though the agent lacks a motivator or a motivating reason at that time.
- When an agent accesses one of her normative reasons, it is capable of serving as a motivating reason.
- In certain conditions, that capacity is realized: a moral reason and motivating reason then converge.
- When this happens, agents earn defeasible moral credit.
- A paradigm of a virtuous agent is one who exemplifies such convergence to a substantial degree.
- A paradigm of a vicious agent is one for whom there is a systematic divergence between his moral reasons, on the one hand, and his motivating reasons, on the other.

This is just the beginning of a more extensive list that could easily be cobbled together.[21] All thirteen of these claims are reason-based, and with respect to the connection between moral reasons and motivation, they are, when taken together, accurate, robust, illuminating, coherent, and orderly. Although they comprise only a partial view of their target, they already jointly realize the six features of theoretical understanding, and so represent substantial noetic progress regarding a foremost question of contemporary metaethics.

Russell's challenge and contemporary philosophy of perception

....

What about Russell's challenge? He rightly suggested that mineralogists and other scientists could cite many truths that would draw the consensus of their peers, while perhaps being news to non-specialists.

---

[21] Note that these claims are not data, as they belong to what we called (in Chapter 2, §3) a 'well-formed' theory of the connection between moral reasons and motivation.

Now consider the body of truths contemporary philosophers of perception could give if asked to do the same for their field. These include:

- Seeing is not believing: perceiving an object as having some feature does not require believing that the object has that feature.
- Perceiving *o*'s being *F* requires more than perceiving *o* and perceiving *F*.
- Perceiving *o*'s being *F* is not identical to knowledge that *o* is *F*, though it can serve as the basis for such knowledge.
- Perception can ground *de re* reference and singular thought.
- Perception is at least sometimes theory-laden.
- Experts perceive more or better than novices at least partly because the former possess a richer conceptual repertoire than the latter.
- Something is a perceptual state only if it is capable of providing information that guides its subject's behavior and affects its subject's non-perceptual representational states.
- Some perceptual states are perceptual experiences, which are individuated by their phenomenal character.
- All perceptual experiences exhibit both phenomenal and access consciousness.
- Successfully perceiving *o* requires more than a veridical perceptual experience as of *o*.
- Two perceptual states can be subjectively indistinguishable yet differ in their phenomenal character.
- Perceptual experiences are corrigible: having a perceptual experience does not require believing that one is having that experience.
- Believing that one is having a certain perceptual experience does not entail that one is having that experience.
- The phenomenal character of perceptual experience is very finely grained.
- The phenomenal character of perceptual experience is non-transitive.

And so on—these are just the tip of a sizeable iceberg.

2.5 The role of the Tri-Level Method

We've seen that the Tri-Level Method plays a central role in answering the how-to and evaluative questions. It is natural to ask whether it sheds light on the progress just described. We believe that it does. By endorsing all thirteen claims regarding the connection between moral reasons and motivation, contemporary meta-ethicists are better positioned than their predecessors to *accommodate* and *explain* data about such reasons. They have also *substantiated* and *integrated* these claims to a greater extent than those asserted by earlier views. In the previous chapter we saw that the method's criteria are constitutively linked to the six features of theoretical understanding. So it should come as no surprise that realization of these features by the set of claims we've listed above correlates with satisfaction of the method's criteria.

Our examples illustrate how the Tri-Level Method underlies noetic progress with respect to the foremost questions in philosophy. They also reveal the method's ability to help explain how such progress has occurred: namely, through success at accommodation, explanation, substantiation, and integration with respect to the subject matters of those questions. To be clear, our claim is not that the construction of views that satisfy the method's criteria has always been *intentional*. Rather, we suspect that in many cases philosophers have intended to satisfy criteria invoked by other methods. Or they have meant simply to engage in one of the activities that typify philosophical practice (arguing, objecting, replying, explaining, etc.). Either way, in so doing, they have managed to satisfy the method's criteria.

The Tri-Level Method also helps to explain why there has been a relative dearth of progress with respect to philosophy's fundamental questions. For philosophers have not constructed well-substantiated views that accommodate and explain most or all of the data regarding the subject matters of the fundamental questions in a manner that is internally and externally integrated. Still, insofar as they have done so with regard to a wide range of the subject matters of the foremost questions, they have developed views that vindicate the claim that there has been noetic progress in philosophy.

Let us take stock. The Tri-Level Method's constitutive link to understanding paves the way for answers to the how-to and evaluative questions. Here we have argued for three further, interrelated conclusions, in reply to the existential and comparative questions.

The first is the rejection of Russell's negative thesis, in conjunction with a diagnosis of its allure—to wit, a failure to distinguish fundamental questions from foremost ones. Our contention is not that philosophy is fully on a par with mathematics and the natural sciences regarding their respective questions. Rather, our point is that this distinction shows the situation to be not nearly as dire as many allege. It would be harmless if philosophy's progress with respect to its *fundamental* questions compared unfavorably with progress in mathematics and the natural sciences with respect to their *foremost* questions. Although comparing apples to oranges in this way is clearly problematic, it regularly impels critiques of philosophical progress.

We've also argued—as our second conclusion—that when it comes to foremost questions, the Tri-Level Method helps to explain philosophical progress by identifying its probable source: philosophers have developed views that satisfy the method's criteria better than their predecessors. Unsurprisingly, they have yet to resolve philosophy's fundamental questions, whose scope and complexity render the task of constructing respectable (let alone excellent) theories regarding these questions extraordinarily difficult. In fact, such questions may be unresolvable; nevertheless, even if there is in this sense no "end" to philosophy, were theorists to satisfy the method's criteria to a high degree, this would secure noetic progress with respect to the subject matters of those questions.

But insofar as such progress is absent with respect to fundamental questions—this is our third conclusion—the Tri-Level Method helps to explain this (and its relative innocence).

Of course, there is a sense in which advancement even with respect to philosophy's foremost questions has been slow and haphazard; while there has been progress on these questions, it has come in fits and starts. But this is only to be expected. For a method whose basic criteria are constitutively linked to theoretical understanding, as the Tri-Level Method's are, has not previously been formulated, much less fully implemented. Here, too, we believe that, with suitable methodological discipline, progress is within reach.

REFERENCES

Adler, Mortimer Jerome. 1927. *Dialectic*. Harcourt, Brace and Company.

Adorno, Theodor W. 1993. *Hegel: Three Studies*. Translated by Shierry Weber Nicholsen. MIT Press.

Anderson, Elizabeth. 1995. "Knowledge, Human Interests, and Objectivity in Feminist Epistemology." *Philosophical Topics*, 23: 27–58.

Anscombe, G. E. M. 1957. *Intention*. Basil Blackwell.

Archer, Avery. 2018. "Wondering about What You Know." *Analysis*, 78: 596–604.

Armstrong, D. M. 1973. "Epistemological Foundations for a Materialistic Theory of the Mind." *Philosophy of Science*, 40: 178–93.

Armstrong, D. M. 1989. *Universals: An Opinionated Introduction*. Westview Press.

Arras, John D. 2007. "The Way We Reason Now: Reflective Equilibrium in Bioethics." In Bonnie Steinbock, ed. *The Oxford Handbook of Bioethics*. Oxford University Press: 46–71.

Austin, J. L. 1956–7/1979. "A Plea for Excuses." *Proceedings of the Aristotelian Society*, 57: 1–30.

Ayer, Alfred J. 1936. *Language, Truth and Logic*. Victor Gollancz Ltd.

Baggini, Julian and Peter S. Fosl. 2010. *The Philosopher's Toolkit: A Compendium of Philosophical Concepts and Methods*, second edition. Wiley-Blackwell.

Baier, Annette. 1985. *Postures of the Mind*. University of Minnesota Press.

Baker, Lynne Rudder. 2000. *Persons and Bodies: A Constitution View*. Cambridge University Press.

Ballantyne, Nathan. 2013. "Knockdown Arguments." *Erkenntnis*, 79: 525–43.

Barnes, Elizabeth. 2016. *The Minority Body*. Oxford University Press.

Baz, Avner. 2017. *The Crisis of Method in Contemporary Analytic Philosophy*. Oxford University Press.

Bealer, George. 1992. "The Incoherence of Empiricism." *Aristotelian Society Supplementary Volume*, 66: 99–138.

Bealer, George. 1993. "Universals." *The Journal of Philosophy*, 60: 5–32.

Bealer, George. 1996a. "On the Possibility of Philosophical Knowledge." *Philosophical Perspectives*, 10: 1–34.

Bealer, George. 1996b. "A Theory of Concepts and Concept Possession." *Philosophical Issues*, 9: 261–301.

Bealer, George. 2007. "Mental Causation." *Philosophical Perspectives*, 21: 23–54.

Beatty, John and Alfred Moore. 2010. "Should We Aim for Consensus?" *Episteme*, 7: 198–214.

Beebee, Helen. 2018. "Philosophical Scepticism and the Aims of Philosophy." *Proceedings of the Aristotelian Society*, 118: 1–24.

Benacerraf, Paul. 1973. "Mathematical Truth." *The Journal of Philosophy*, 70: 661–79.

Bengson, John. 2015. "Grasping the Third Realm." *Oxford Studies in Epistemology*, 5: 1–38.

Bengson, John. 2017. "The Unity of Understanding." In Stephen Grimm, ed. *Making Sense of the World: New Essays on the Philosophy of Understanding*. Oxford University Press: 14–53.

Bengson, John. 2020. "*True Enough*." *Mind*, 129: 256–68.

Bengson, John and Marc A. Moffett. 2011. "Non-Propositional Intellectualism." In John Bengson and Marc A. Moffett, eds. *Knowing How: Essays on Knowledge, Mind, and Action*. Oxford University Press: 161–95.

Bengson, John, Terence Cuneo, and Russ Shafer-Landau. 2019a. "Methods, Goals, and Data in Moral Theorizing." In Karen Jones, Mark Timmons, and Aaron Zimmerman, eds. *The Routledge Handbook of Moral Epistemology*. Routledge: 401–18.

Bengson, John, Terence Cuneo, and Russ Shafer-Landau. 2019b. "Method in the Service of Progress." *Analytic Philosophy*, 60: 179–205.

Bengson, John, Terence Cuneo, and Russ Shafer-Landau. 2023. *The Moral Universe*. Oxford University Press.

Bengson, John, Terence Cuneo, and Russ Shafer-Landau. Forthcoming. *Grasping Morality*. Oxford University Press.

Bengson, John, Marc A. Moffett, and Jennifer C. Wright. 2009. "The Folk on Knowing How." *Philosophical Studies*, 142: 387–401.

Bird, Alexander. 2007. "What Is Scientific Progress?" *Noûs*, 41: 64–89.

Bird, Otto. 1953. "How to Read an Article of the *Summa*." *New Scholasticism*, 27: 129–59.

Blackburn, Simon. 1993. *Essays in Quasi-Realism*. Oxford University Press.

Blackburn, Simon. 1998. *Ruling Passions: A Theory of Practical Reasoning*. Clarendon Press.

Block, Ned and Jerry A. Fodor. 1972. "What Psychological States Are Not." *The Philosophical Review*, 81: 159–81.

Boghossian, Paul. 2006. *Fear of Knowledge: Against Relativism and Constructivism*. Oxford University Press.

Bonevac, Daniel. 2004. "Reflection Without Equilibrium." *The Journal of Philosophy*, 101: 363–88.

Bonevac, Daniel, Josh Dever, and David Sosa. 2011. "The Counterexample Fallacy." *Mind*, 120: 1143–58.

BonJour, Laurence. 1997. *In Defense of Pure Reason: A Rationalist Account of A Priori Justification*. Cambridge University Press.

BonJour, Laurence. 2014. "In Defense of the A Priori." In Matthias Steup, John Turri, and Ernest Sosa, eds. *Contemporary Debates in Epistemology*, second edition. Wiley-Blackwell: 177–84.

Brandt, Richard. 1979. *A Theory of the Good and the Right*. Oxford University Press.

Brandt, Richard. 1990. "The Science of Man and Wide Reflective Equilibrium." *Ethics*, 100: 259–78.

Bright, Liam Kofi. 2017. "On Fraud." *Philosophical Studies*, 174: 291–310.

Brink, David. 1989. *Moral Realism and the Foundations of Ethics*. Cambridge University Press.

Bronstein, David. 2016. *Aristotle on Knowledge and Learning: The Posterior Analytics*. Oxford University Press.

Burgess, Alexi, Herman Cappelen, and David Plunkett, eds. 2020. *Conceptual Engineering and Conceptual Ethics*. Oxford University Press.

Burgess, John and Gideon Rosen. 1997. *A Subject with No Object: Strategies for Nominalistic Interpretations of Mathematics*. Oxford University Press.

Burnyeat, Myles. 1981. "Aristotle on Understanding Knowledge." In Enrico Berti, ed. *Aristotle on Science: The Posterior Analytics*. Edizioni Antenori: 97–139.

Cappelen, Herman. 2018. *Fixing Language: An Essay on Conceptual Engineering*. Oxford University Press.

Cappelen, Herman, Tamar Szabó Gendler, and John Hawthorne, eds. 2016. *The Oxford Handbook of Philosophical Methodology*. Oxford University Press.

Carmichael, Chad. 2010. "Universals." *Philosophical Studies*, 150: 373–89.

Carnap, Rudolf. 1950/1962. *Logical Foundations of Probability*, second edition. University of Chicago Press.

Castañeda, Hector-Neri. 1980. *On Philosophical Method*. Noûs Publications.

Cath, Yuri. 2011. "Knowing How Without Knowing That." In John Bengson and Marc A. Moffett, eds. *Knowing How: Essays on Knowledge, Mind, and Action*. Oxford University Press: 113–35.

Chalmers, David. 2002. "Does Conceivability Entail Possibility?" In Tamar Szabó Gendler and John Hawthorne, eds. *Conceivability and Possibility*. Oxford University Press: 145–200.

Chalmers, David. 2014. "Why Isn't There More Progress in Philosophy?" *Philosophy*, 90: 3–31. Reprinted in Ted Honderich, ed. 2015. *Philosophers of Our Times*. Oxford University Press: 347–70.

Chalmers, David, David Manley, and Ryan Wasserman, eds. 2009. *Metametaphysics: New Essays on the Foundations of Ontology*. Oxford University Press.

Chisholm, Roderick M. 1977. *Theory of Knowledge*, second edition. Prentice Hall.

Churchland, Paul M. 1984. *Matter and Consciousness*. MIT Press.

Clatterbuck, Hayley. 2020. "A Defense of Low-Probability Scientific Explanations." *Philosophy of Science*, 87: 91–112.

Collingwood, R. G. 1933. *An Essay on Philosophical Method*. Oxford University Press.

Copp, David. 1985. "Morality, Reason, and Management Science: The Rationale of Cost-Benefit Analysis." *Social Philosophy and Policy*, 2: 128–51.

Craig, Edward. 1990. *Knowledge and the State of Nature: An Essay in Conceptual Synthesis*. Oxford University Press.

Cummins, Robert. 1998. "Reflection on Reflective Equilibrium." In Michael DePaul and William Ramsey, eds. *Rethinking Intuition*. Rowman & Littlefield Publishers, Inc.: 113–28.

Daly, Christopher. 2010. *An Introduction to Philosophical Method*. Broadview Press.

Daly, Christopher, ed. 2015. *The Palgrave Handbook of Philosophical Methods*. Palgrave Macmillan.

Dang, Haixin. 2019. "Do Collaborators in Science Need to Agree?" *Philosophy of Science*, 86: 1029–40.

Daniels, Norman. 1979. "Wide Reflective Equilibrium and Theory Acceptance in Ethics." *The Journal of Philosophy*, 76: 256–82.

Daniels, Norman. 1996. *Justice and Justification: Reflective Equilibrium in Theory and Practice*. Cambridge University Press.

Deleuze, Gilles and Félix Guattari. 1996. *What Is Philosophy?* Columbia University Press.

Dellsén, Finnur. 2016. "Scientific Progress: Knowledge versus Understanding." *Studies in History and Philosophy of Science*, 56: 72–83.

Dennett, Daniel C. 2013. *Intuition Pumps and Other Tools for Thinking.* W. W. Norton Company.

DePaul, Michael. 1993. *Balance and Refinement: Beyond Coherence Methods of Moral Inquiry.* Routledge.

DePaul, Michael. 1998. "Why Bother with Reflective Equilibrium?" In Michael DePaul and William Ramsey, eds. *Rethinking Intuition.* Rowman & Littlefield Publishers, Inc.: 293–309.

Derrida, Jacques. 1991. "Letter to a Japanese Friend." In Peggy Kamuf, ed. *A Derrida Reader.* Harvester: 270–6.

Descartes, René. 1628/1985. *Rules for the Direction of the Mind.* Translated by John Cottingham, Robert Stoothoff, and Dugald Murdoch. *The Philosophical Writings of Descartes*, vol. 1. Cambridge University Press.

Devitt, Michael. 2014. "There Is No A Priori." In Matthias Steup, John Turri, and Ernest Sosa, eds. *Contemporary Debates in Epistemology*, second edition. Wiley-Blackwell: 185–94.

Dewey, John. 1910. *How We Think.* D. C. Heath.

Dewey, John. 1938. *Logic: The Theory of Inquiry.* Henry Holt.

Dietrich, Eric. 2010. "There Is No Progress in Philosophy." *Essays in Philosophy*, 12: 329–44.

Doris, John. 2002. *Lack of Character: Personality and Moral Behavior.* Cambridge University Press.

D'Oro, Giuseppina and Søren Overgaard, eds. 2017. *The Cambridge Companion to Philosophical Methodology.* Cambridge University Press.

Dorr, Cian. 2007. "There Are No Abstract Objects." In Theodore Sider, John Hawthorne, and Dean W. Zimmerman, eds. *Contemporary Debates in Metaphysics.* Blackwell: 32–64.

Doyle, Arthur Conan. 1887. *A Study in Scarlet: A Sherlock Holmes Murder Mystery.* Ward, Lock, and Co.

Dreyfus, Hubert L., ed. 2014. *Skillful Coping: Essays on the Phenomenology of Everyday Perception and Action.* Oxford University Press.

Dubois, W. E. B. 1898. "The Study of Negro Problems." *The Annals of the American Academy of Political and Social Science*, 11: 1–23.

Duhem, Pierre. 1914. *La Théorie Physique: Son Objet et sa Structure*. Marcel Riviera & Cie.

Dummett, Michael. 1978. "Can Analytical Philosophy Be Systematic, and Ought It to Be?" In Michael Dummett, ed. *Truth and Other Enigmas*. Harvard University Press: 437–58.

Edwards, Douglas. 2018. *The Metaphysics of Truth*. Oxford University Press.

Elga, Adam. 2007. "Reflection and Disagreement." *Noûs*, 41: 479–502.

Elgin, Catherine Z. 1996. *Considered Judgment*. Princeton University Press.

Elgin, Catherine Z. 2017. *True Enough*. MIT Press.

Enoch, David. 2011. *Taking Morality Seriously*. Oxford University Press.

Enoch, David. 2019. "Non-Naturalistic Realism in Metaethics." In Tristram McPherson and David Plunkett, eds. *The Routledge Handbook of Metaethics*. Routledge: 29–42.

Feldman, Fred. 1992. *Confrontations with the Reaper: A Philosophical Study of the Nature and Value of Death*. Oxford University Press.

Feldman, Richard. 2007. "Reasonable Religious Disagreements." In Louise Antony, ed. *Philosophers Without Gods*. Oxford University Press: 194–214.

Feyerabend, Paul. 1975. *Against Method*. New Left Books.

Field, Hartry. 1989. *Realism, Mathematics and Modality*. Blackwell.

Field, Hartry. 2005. "Recent Debates about the A Priori." *Oxford Studies in Epistemology*, 1: 69–88.

Finlay, Stephen. 2014. *Confusion of Tongues: A Theory of Normative Language*. Oxford University Press.

Finn, Suki. 2018. "Methodology for the Metaphysics of Pregnancy." Presented at the Logic and Metaphysics Workshop of the CUNY Graduate Center, New York.

Frances, Bryan. 2015. "Worrisome Skepticism about Philosophy." *Episteme*, 13: 289–303.

Friedman, Jane. 2017. "Why Suspend Judging?" *Noûs*, 51: 302–26.

Friedman, Jane. 2019. "Inquiry and Belief." *Noûs*, 53: 296–315.

Gadamer, Hans Georg. 1960/1989. *Truth and Method*, second revised edition. Bloomsbury Press.

Gendler, Tamar Szabó and John Hawthorne, eds. 2002. *Conceivability and Possibility*. Oxford University Press.

Gibbard, Allan. 2003. *Thinking How to Live*. Harvard University Press.

Glick, Ephraim. 2011. "Two Methodologies for Evaluating Intellectualism." *Philosophy and Phenomenological Research*, 83: 398–434.

Godfrey-Smith, Peter. 2003. *Theory and Reality: An Introduction to the Philosophy of Science*. University of Chicago Press.

Godfrey-Smith, Peter. 2006. "Theories and Models in Metaphysics." *The Harvard Review of Philosophy*, 14: 4–19.

Goldberg, Sanford. 2013. "Disagreement, Defeaters, and Assertion." In David Christensen and Jennifer Lackey, eds. *The Epistemology of Disagreement: New Essays*. Oxford University Press.

Goldman, Alvin I. 1970. *A Theory of Human Action*. Princeton University Press.

Goldwater, Jonah. 2021. "The Lump and the Ledger: Material Coincidence at Little-to-No Cost." *Erkenntnis*, 86: 789–812.

Goodman, Nelson. 1955. *Fact, Fiction, and Forecast*. Harvard University Press.

Gozzano, Simone. 2006. "Functional Role Semantics and Reflective Equilibrium." *Acta Analytica*, 21: 62–72.

Gozzano, Simone and Christopher S. Hill, eds. 2015. *New Perspectives on Type Identity: The Mental and the Physical*. Cambridge University Press.

Gutting, Gary. 2009. *What Philosophers Know*. Cambridge University Press.

Gutting, Gary. 2016. "Philosophical Progress." In Herman Cappelen, Tamar Szabó Gendler, and John Hawthorne, eds. *Oxford Handbook of Philosophical Methodology*. Oxford University Press: 309–25.

Haack, Susan. 1993. *Evidence and Inquiry*. Wiley-Blackwell.

Hadot, Pierre. 1981/1995. *Philosophy as a Way of Life*. Wiley-Blackwell.

Hájek, Alan. 2016. "Philosophical Heuristics and Philosophical Methodology." In Herman Cappelen, Tamar Szabó Gendler, and John Hawthorne, eds. *The Oxford Handbook of Philosophical Methodology*. Oxford University Press: 348–73.

Hanson, Norwood Russell. 1958. *Patterns of Discovery: An Inquiry into the Conceptual Foundations of Science*. Cambridge University Press.

Hardimon, Michael O. 2017. *Rethinking Race: The Case for Deflationary Realism*. Harvard University Press.

Hare, R. M. 1973. "Rawls' Theory of Justice." *Philosophical Quarterly*, 23: 144–55; 241–51.

Harman, Gilbert. 1977. *The Nature of Morality*. Oxford University Press.

Harman, Gilbert. 1999. "Moral Philosophy Meets Social Psychology: Virtue Ethics and the Fundamental Attribution Error." *Proceedings of the Aristotelian Society*, 99: 315–31.

Harman, Gilbert. 2010. "Epistemology as Methodology." In Jonathan Dancy, Ernest Sosa, and Matthias Steup, eds. *A Companion to Epistemology*, second edition. Wiley-Blackwell: 152–6.

Haslanger, Sally. 2012. *Resisting Reality: Social Construction and Social Critique*. Oxford University Press.

Haug, Matthew C., ed. 2013. *Philosophical Methodology: The Armchair or the Laboratory?* Routledge.

Heidegger, Martin. 1927/2008. *Being and Time*. Translated by John MacQuarrie and Edward Robinson. Harper Perennial Modern Classics.

Heney, Diana B. 2016. *Towards a Pragmatist Metaethics*. Routledge.

Hilbert, David. 1902. "Mathematical Problems." *Bulletin of the American Mathematical Society*, 8: 437–79.

Hintikka, Jaakko. 1999. *Inquiry as Inquiry: A Logic of Scientific Discovery*. Springer.

hooks, bell. 1991. "Theory as Liberatory Practice." *Yale Journal of Law and Feminism*, 4: 1–12.

Horwich, Paul. 2012. *Wittgenstein's Metaphilosophy*. Oxford University Press.

Huemer, Michael. 2008. "Revisionary Intuitionism." *Social Philosophy and Policy*, 25: 368–92.

Huemer, Michael. 2009. "When Is Parsimony a Virtue?" *The Philosophical Quarterly*, 59: 216–36.

Jackson, Frank. 1982. "Epiphenomenal Qualia." *Philosophical Quarterly*, 32: 127–36.

Jackson, Frank. 1998. *From Metaphysics to Ethics: A Defense of Conceptual Analysis*. Oxford University Press.

Jaggar, Alison. 1989. "Love and Knowledge: Emotion in Feminist Epistemology." *Inquiry*, 32: 151–76.

Kamm, F. M. 1993. *Morality, Mortality, vol. 1: Death and Whom to Save from It*. Oxford University Press.

Kamm, F. M. 1996. *Morality, Mortality, vol. 2: Rights, Duties, and Status*. Oxford University Press.

Kamm, F. M. 2007. *Intricate Ethics: Rights, Responsibilities, and Permissible Harms*. Oxford University Press.

Kamtekar, Rachana. 2004. "Situationism and Virtue Ethics on the Content of Our Character." *Ethics*, 114: 458–91.

Kant, Immanuel. 1787/1998. *Critique of Pure Reason*. Translated and edited by Paul Guyer and Allen Wood. Cambridge University Press.

Keefe, Rosanna. 1999. "Introduction: Theories of Vagueness." In Rosanna Keefe and Peter Smith, eds. *Vagueness: A Reader*. Cambridge University Press: 1–57.

Kelly, Thomas and Sarah McGrath. 2010. "Is Reflective Equilibrium Enough?" *Philosophical Perspectives*, 24: 325–59.

Kelp, Christoph. 2021. "Theory of Inquiry." *Philosophy and Phenomenological Research*, 103: 359–384.

Khader, Serene. 2018. *Decolonizing Universalism: A Transnational Feminist Ethic.* Oxford University Press.

Kitcher, Philip. 1981. "Explanatory Unification." *Philosophy of Science*, 48: 507–31.

Kitcher, Philip. 1990. "The Division of Cognitive Labour." *The Journal of Philosophy*, 87: 5–22.

Kitcher, Philip. 1993. *The Advancement of Science: Science Without Legend, Objectivity Without Illusions.* Oxford University Press.

Kitcher, Philip. 2000. "A Priori Knowledge Revisited." In Paul Boghossian and Christopher Peacocke, eds. *New Essays on the A Priori.* Oxford University Press: 65–91.

Korman, Daniel Z. 2015. *Objects: Nothing Out of the Ordinary.* Oxford University Press.

Kornblith, Hilary. 2013. "Is Philosophical Knowledge Possible?" In Diego E. Machuca, ed. *Disagreement and Skepticism.* Routledge: 260–77.

Korsgaard, Christine. 1996. *The Sources of Normativity.* Harvard University Press.

Kriegel, Uriah. 2013. "The Epistemological Challenge for Revisionary Metaphysics." *Philosophers' Imprint*, 13: 1–30.

Kuhn, Thomas. 1962. *The Structure of Scientific Revolutions.* University of Chicago Press.

Kvanvig, Jonathan. 2003. *The Value of Knowledge and the Pursuit of Understanding.* Cambridge University Press.

Lerner, Adam and Sarah-Jane Leslie. 2013. "Generics, Generalism, and Reflective Equilibrium: Implications for Moral Theorizing from the Study of Language." *Philosophical Perspectives*, 27: 366–403.

Levine, Joseph. 2001. *Purple Haze: The Puzzle of Consciousness.* Oxford University Press.

Lewis, David. 1983. *Philosophical Papers.* Oxford University Press.

Lewis, David. 1986. *On the Plurality of Worlds.* Basil Blackwell.

Lewis, David and Stephanie Lewis. 1970. "Holes." *Australasian Journal of Philosophy*, 48: 206–12.

Linnebo, Øystein. 2006. "Epistemological Challenges to Mathematical Platonism." *Philosophical Studies*, 129: 545–74.

Livengood, Jonathan and Justin Sytsma. 2015. *The Theory and Practice of Experimental Philosophy*. Broadview Press.

Lycan, William. 2019. *On Evidence in Philosophy*. Oxford University Press.

Machery, Edouard. 2017. *Philosophy Within Its Proper Bounds*. Oxford University Press.

MacIntyre, Alasdair. 1988. *Whose Justice? Which Rationality?* University of Notre Dame Press.

Maitra, Ishani. 2016. "Feminism." In Herman Cappelen, Tamar Szabó Gendler, and John Hawthorne, eds. *The Oxford Handbook of Philosophical Methodology*. Oxford University Press: 690–708.

Malebranche, Nicolas. 1674/1997. *The Search after Truth*. Translated by Thomas M. Lennon and Paul J. Olscamp. Cambridge University Press.

Manne, Kate. 2017. *Down Girl: The Logic of Misogyny*. Oxford University Press.

Massimi, Michaela. 2018. "Four Kinds of Perspectival Truth." *Philosophy and Phenomenological Research*, 96: 342–59.

McGinn, Colin. 1993. *Problems in Philosophy: The Limits of Inquiry*. Basil Blackwell.

McGrath, Sarah. 2019. *Moral Knowledge*. Oxford University Press.

McNaughton, David. 1988. *Moral Vision: An Introduction to Ethics*. Blackwell.

McPherson, Tristram. 2015. "The Methodological Irrelevance of Reflective Equilibrium." In Christopher Daly, ed. *The Palgrave Handbook of Philosophical Methods*. Palgrave Macmillan: 652–74.

McPherson, Tristram. 2020. *Epistemology and Methodology in Ethics*. Cambridge University Press.

Melnyk, Andrew. 2003. *A Physicalist Manifesto: Thoroughly Modern Materialism*. Cambridge University Press.

Mill, J. S. 1867/1900. *A System of Logic, Ratiocinative and Inductive, Being a Connected View of the Principles of Evidence and the Methods of Scientific Investigation*. Longmans, Green, and Co.

Mirrachi, Lisa. 2015. "Knowledge Is All You Need." *Philosophical Issues*, 25: 353–78.

Moore, G. E. 1903. *Principia Ethica*. Cambridge University Press.

Mueller, Gustav Emil. 1965. *Plato, the Founder of Philosophy as Dialectic*. Philosophical Library.

Nado, Jennifer. 2019. "Knowledge Is Not Enough." *Australasian Journal of Philosophy*, 95: 658–72.

Nagel, Thomas. 1986. *The View from Nowhere*. Oxford University Press.

Niinuluoto, Ilka. 1984. *Is Science Progressive?* Reidel.

Nolan, Daniel. 2015. "Lewis's Philosophical Method." In Barry Loewer and Jonathan Schaffer, eds. *Blackwell Companion to David Lewis*. Wiley-Blackwell.

Nozick, Robert. 1981. *Philosophical Explanations*. Harvard University Press.

Olson, Jonas. 2014. *Moral Error Theory*. Oxford University Press.

Overgaard, Søren, Paul Gilbert, and Stephen Burwood. 2013. *An Introduction to Metaphilosophy*. Cambridge University Press.

Papineau, David. 2012. *Philosophical Devices: Proofs, Probabilities, Possibilities, and Sets*. Oxford University Press.

Papineau, David. 2017. "Is Philosophy Simply Harder Than Science?" *Times Literary Supplement Online* (June 1): www.the-tls.co.uk/articles/public/philosophy-simply-harder-science.

Parfit, Derek. 1984. *Reasons and Persons*. Oxford University Press.

Parfit, Derek. 2011. *On What Matters*, volume 2. Oxford University Press.

Paul, L. A. 2012. "Metaphysics as Modeling: The Handmaiden's Tale." *Philosophical Studies*, 160: 1–29.

Peacocke, Christopher. 2004. *The Realm of Reason*. Oxford University Press.

Peirce, Charles Sanders. 1877. "The Fixation of Belief." *Popular Science Monthly*, 12: 1–15.

Peregrin, Jaroslav and Vladimír Svoboda. 2017. *Reflective Equilibrium and the Principles of Logical Analysis: Understanding the Laws of Logic*. Routledge.

Plunkett, David. 2015. "Which Concepts Should We Use? Metalinguistic Negotiations and the Methodology of Philosophy." *Inquiry*, 58: 828–74.

Price, H. H. 1945. "Clarity Is Not Enough." *Aristotelian Society Supplementary Volume*, 19: 1–31.

Priest, Graham. 2006. "What Is Philosophy?" *Philosophy*, 81: 189–207.

Pritchard, Duncan. 2010. "Knowledge and Understanding." In *The Nature and Value of Knowledge: Three Investigations*, co-authored with Alan Millar and Adrian Haddock. Oxford University Press: 3–88.

Pritchard, Duncan. 2016. "Seeing It for Oneself: Perceptual Knowledge, Understanding, and Intellectual Autonomy." *Episteme*, 13: 29–42.

Pust, Joel. 2000. *Intuition as Evidence*. Routledge.

Putnam, Hilary. 1967. "Psychological Predicates." In W. H. Capitan and D. D. Merrill, eds. *Art, Mind, and Religion*. University of Pittsburgh Press: 37–48.

Quine, W. V. O. 1948. "On What There Is." *Review of Metaphysics*, 2: 21–38.

Quine, W. V. O. 1951. "Two Dogmas of Empiricism." *The Philosophical Review*, 60: 20–43.

Quine, W. V. O. 1970. "Reasons for the Indeterminacy of Translation." *The Journal of Philosophy*, 67: 178–83.

Quine, W. V. O. and J. S. Ullian. 1970. *The Web of Belief*. Random House.

Ragland, C. P. and Sarah Heidt, eds. 2001. *What Is Philosophy?* Yale University Press.

Rawls, John. 1971. *A Theory of Justice*. Harvard University Press.

Rawls, John. 1974. "The Independence of Moral Theory." *Proceedings and Addresses of the American Philosophical Association*, 47: 5–22.

Rawls, John. 1980. "Kantian Constructivism in Moral Theory." *The Journal of Philosophy*, 77: 515–72.

Rayo, Agustin. 2013. *The Construction of Logical Space*. Oxford University Press.

Raz, Joseph. 1982. "The Claims of Reflective Equilibrium." *Inquiry*, 25: 307–30.

Rescher, Nicholas. 2006. *Philosophical Dialectics: An Essay on Metaphilosophy*. State University of New York Press.

Rorty, Richard, ed. 1967. *The Linguistic Turn*. University of Chicago Press.

Rorty, Richard. 1979. *Philosophy and the Mirror of Nature*. Princeton University Press.

Rosen, Michael. 1982. *Hegel's Dialectic and Its Criticism*. Cambridge University Press.

Rumfitt, Ian. 2003. "Savoir Faire." *The Journal of Philosophy*, 100: 158–66.

Russell, Bertrand. 1912/1997. *The Problems of Philosophy*. Oxford University Press.

Russell, Bertrand. 1914. *Our Knowledge of the External World as a Field for Scientific Method in Philosophy*. George Allen and Unwin Limited.

Ryle, Gilbert. 1949. *The Concept of Mind*. Hutchinson & Co.

Sayre-McCord, Geoffrey. 1996. "Coherentist Epistemology and Moral Theory." In Walter Sinnott-Armstrong and Mark Timmons, eds. *Moral Knowledge*. Oxford University Press: 137–89.

Scanlon, T. M. 2002. "Rawls on Justification." In Samuel Freeman, ed. *The Cambridge Companion to Rawls*. Cambridge University Press: 139–67.

Scanlon, T. M. 2014. *Being Realistic about Reasons*. Oxford University Press.

Sellars, Wilfrid R. 1956. "Empiricism and the Philosophy of Mind." *Minnesota Studies in the Philosophy of Science*, 1: 253–329.

Shafer-Landau, Russ. 2003. *Moral Realism: A Defense*. Oxford University Press.

Shapiro, Lawrence A. 2000. "Multiple Realizations." *The Journal of Philosophy*, 97: 635–54.

Sider, Theodore. 2013. "Against Parthood." *Oxford Studies in Metaphysics*, 8: 237–93.

Sider, Theodore. 2020. *The Tools of Metaphysics and the Metaphysics of Science*. Oxford University Press.

Siegel, Susanna. 2010. *The Contents of Visual Experience*. Oxford University Press.

Siegel, Susanna. 2017. *The Rationality of Perception*. Oxford University Press.

Smith, Nicholas. 2020. "Simply Finding Answers, or the Entirety of Inquiry while Standing on One Foot." *Disputatio*, 12: 181–98.

Sober, Elliott. 2009. "Parsimony Arguments in Science and Philosophy: A Test Case for Naturalism." *Proceedings and Addresses of the American Philosophical Association*, 83: 117–55.

Solomon, Miriam. 2001. *Social Empiricism*. MIT Press.

Sosa, Ernest. 2019. "Suspension as Spandrel." *Episteme*, 16: 357–68.

Sreenivasan, Gopal. 2002. "Errors about Errors: Virtue Theory and Trait Attribution." *Mind*, 111: 47–68.

Stalnaker, Robert. 1984. *Inquiry*. Cambridge University Press.

Stanley, Jason and Timothy Williamson. 2001. "Knowing How." *The Journal of Philosophy*, 98: 411–44.

Stebbing, L. S. 1932–3. "The Method of Analysis in Metaphysics." *Proceedings of the Aristotelian Society*, 33: 65–94.

Stich, Stephen. 1988. "Reflective Equilibrium, Analytic Epistemology and the Problem of Cognitive Diversity." *Synthese*, 74: 391–413.

Stoljar, Daniel. 2017. *Philosophical Progress: In Defence of a Reasonable Optimism*. Oxford University Press.

Strawson, Peter. 1992. *Analysis and Metaphysics*. Oxford University Press.

Street, Sharon. 2008. "Constructivism about Reasons." *Oxford Studies in Metaethics*, 3: 207–46.

Street, Sharon. 2015. "Does Anything Really Matter or Did We Just Evolve to Think So?" In Alex Byrne, Joshua Cohen, Gideon Rosen, and Seana Shiffrin, eds. *The Norton Introduction to Philosophy*. Norton: 685–93.

Thomson, Judith Jarvis. 1990. *The Realm of Rights*. Harvard University Press.

Trout, J. D. 2002. "Scientific Explanation and the Sense of Understanding." *Philosophy of Science*, 69: 212–33.

Urmson, J. O. 1967. "J. L. Austin." In Richard Rorty, ed. *The Linguistic Turn*. University of Chicago Press: 232–8.

van Inwagen, Peter. 2006. *The Problem of Evil*. Oxford University Press.

Walden, Kenneth. 2013. "In Defense of Reflective Equilibrium." *Philosophical Studies*, 166: 243–56.

Weatherson, Brian. 2003. "What Good Are Counterexamples?" *Philosophical Studies*, 115: 1–31.

Weinberg, Jonathan M. 2007. "How to Challenge Intuitions Empirically Without Risking Skepticism." *Midwest Studies in Philosophy*, 31: 318–43.

Whitcomb, Dennis. 2010. "Curiosity Was Framed." *Philosophy and Phenomenological Research*, 81: 664–87.

Wiggins, David. 2001. *Sameness and Substance Renewed*. Oxford University Press.

Williams, Bernard. 1985. *Ethics and the Limits of Philosophy*. Harvard University Press.

Williamson, Timothy. 2000. *Knowledge and Its Limits*. Oxford University Press.

Williamson, Timothy. 2007. *The Philosophy of Philosophy*. Blackwell.

Williamson, Timothy. 2013. *Modal Logic as Metaphysics*. Oxford University Press.

Williamson, Timothy. 2018. *Doing Philosophy: From Common Curiosity to Logical Reasoning*. Oxford University Press. Reprinted as *Philosophical Method: A Very Short Introduction*. 2020. Oxford University Press.

Wilson, Jessica. 2014. "Three Dogmas of Metaphysical Methodology." In Matthew C. Haug, ed. *Philosophical Methodology: The Armchair or the Laboratory?* Routledge: 145–65.

Wilson, Jessica. 2017. "Three Barriers to Philosophical Progress." In Russell Blackford and Damien Broderick, eds. *Philosophy's Future: The Problem of Philosophical Progress*. Wiley-Blackwell: 91–104.

Wittgenstein, Ludwig. 1953. *Philosophical Investigations*. G. E. M. Anscombe and Rush Rhees, eds. Blackwell.

Wittgenstein, Ludwig. 1969. *On Certainty*. G. E. M. Anscombe and G. H. von Wright, eds. Harper & Row.

Wolchover, Natalie. 2020. "What Is a Particle?" *Quanta* (November 12): www.quantamagazine.org/what-is-a-particle-20201112.

Wright, Crispin. 2004. "Warrant for Nothing (and Foundations for Free)?" *Proceedings of the Aristotelian Society*, 78: 167–212.

Yablo, Stephen. 1992. "Mental Causation." *The Philosophical Review*, 101: 245–80.

Zagzebski, Linda. 2001. "Recovering Understanding." In Matthias Steup, ed. *Knowledge, Truth, and Duty: Essays on Epistemic Justification, Responsibility, and Virtue*. Oxford University Press: 235–51.

Zangwill, Nick. 1992. "Variable Realization: Not Proven." *Philosophical Quarterly*, 42: 214–19.

INDEX